TO SIRI WITH LOVE

ALSO BY JUDITH NEWMAN

You Make Me Feel Like an Unnatural Woman: Diary of a New (Older) Mother

Parents from Hell: Unexpurgated Tales of Good Intentions Gone Awry

Tell Me Another One: A Woman's Guide to Men's Classic Lines

TO SIRI WITH LOVE

A Mother, Her Autistic Son, and the Kindness of Machines

JUDITH NEWMAN

HARPER

NEW YORK . LONDON . TORONTO . SYDNEY

HARPER

A hardcover edition of this book was published in 2017 by HarperCollins Publishers.

HarperCollins books may be purchased for educational, business, or sales promotional use. For information, please email the Special Markets Department at SPsales@harpercollins.com.

Chapter Ten, "To Siri With Love," originally appeared in the *New York Times* on October 17, 2014.

FIRST HARPER PAPERBACKS EDITION PUBLISHED IN 2018.

Designed by Leah Carlson-Stanisic

Library of Congress Cataloging-in-Publication Data has been applied for.

ISBN 978-0-06-241363-5 (pbk.)

HB 08.12.2020

For as he thinketh in his heart, so is he.

—PROVERBS 23:7

"Look, Mommy, that bus says, 'Make a Wish.' What's your wish?"

"My wish is for your happiness, health, and safety your whole life, Gus. What's *your* wish?"

"My wish is to live in New York City my whole life, and to be a really friendly guy."

"But not *too* friendly, right?"

"What's 'too friendly'?"

CONTENTS

AUTHOR'S NOTE

These days, it's considered politically incorrect to call a person "autistic." If you do, you are defining that person entirely by his or her disability. Instead, you're supposed to use "person-first" language: a man with autism, a woman with autism.

I understand the thinking here. It's like calling a person who has dwarfism a dwarf; is short stature the one thing that defines him or her as a human being? But as a tiny friend said to me recently, "Oh, for God's sakes, just call me a dwarf. It's the first thing you see, and *I* know it's not the only thing about me that's interesting."

"Person with autism" also suggests that autism is something bad that one needs distance from. You'd never say "a person with left-handedness" or "a person with Jewishness." Then again, you might say "a person with cancer."

Autism does not entirely define my son, but it informs so much about him and our life together. Saying autism is something you are "with" suggests it's something you carry around and can drop at will, like a purse. There's also something about this pseudo delicacy that is patronizing as hell. Not that I'm against every carefully considered word for disability. I just want the words to be true. When you call autistic people "neurodivergent," for example, you are not stretching for political correctness; you are accurately describing their condition.

(I am all for finding new language that is descriptive and fun and to the point. If you want to ask delicately if someone is autistic, how about asking if he or she is a FOT—Friend of Trains? I mean, if there can be Friends of Dorothy . . .)

There's yet another reason I use the term "autistic" freely here: *You* try writing an entire book using the phrase "person with autism" over and over again. It's clunky. So I will refer to men and women here as autistic. I may also defer to the masculine pronoun when I am talking about people in generalities. There is a significant population of people, both neurotypical and autistic, who do not see gender as binary, and resent the use of "he" or "she." I am happy to call a person anything they want to be called, including "they." But I'm not changing the rules of grammar in a memoir. I mention this because a friend just wrote an excellent book on parenting using the pronoun "they" instead of "he or she" at the insistence of her teenage daughter. I read her book simultaneously loving her parenting philosophy and wanting to punch her in the face.

But whatever her crimes against the English language, my friend consulted in detail with her kids about their place in her book. I did not. This is both selfish and necessary. While Gus's good-natured attitude about being my subject has always been, "I will be a celebrity, Mommy," Henry's feelings change with his mood. At first, he was cavalier in a way guaranteed to irritate me: "Oh, it doesn't matter, I never read what you write anyway." At that point there was talk about whether he would get a share of the profits, and a quick glance at his computer's Google history revealed the phrase "film option." Then, when it dawned on him that people he knows actually do read, he panicked a little. So writing about him became a negotiation, which sometimes he won and sometimes he lost. He was fine about being portrayed as a smart-ass bro. What he worried about was that I might reveal him to be the sweet, caring, wonderful soon-to-be man he really is. I only slipped a little.

There are many good, some great, books about autism (see Resources, page 223)—about the science, the history, treatments. While this book touches on all these subjects, you will never see a review of it that begins, "If you have to read one book about autism . . ." This is not that One Book. It is a slice of life for one family, one kid. But I hope it seems sort of a slice of your life, too.

When I wrote the original story about Gus and Siri for the *New York Times,* Gus was twelve. Most of this book takes place when my sons are thirteen and fourteen. When the publisher tears the manuscript from my hands, they will be fifteen, then sixteen when it hits the remainder bin. Kids grow older, and I wanted this book to be accurate, so I would still be writing and changing it if my agent hadn't called me one day and said, "For the love of God, just put a period on it."

So I did.

INTRODUCTION

My kids and I are at the supermarket.

"*We need turkey and ham!*"

Gus tends to speak in exclamation points. "A half pound! And . . . what, Mommy?" I'm stage-whispering directions, trying to keep the conversation focused on deli meats. Behind the counter, Otto politely slices and listens, occasionally interjecting questions. We're on track here.

And then . . . we're not.

"So! My daddy has been in London for ten days, and he comes back in four days, on Wednesday. He comes in to JFK Airport on American Airlines Flight 100, at Terminal Eight," Gus says, warming to the subject. "What? Yes, Mom says thin slices. Also, coleslaw! Daddy will take the A train from Howard Beach to West 4th and then change to the B or D to Broadway–Lafayette. He'll arrive at 77 Bleecker in the morning and then he and Mommy will do sex . . ."

Suddenly Otto is interested. "What?"

"You know, my daddy? The one who is old and has bad knees? He arrives at Terminal Eight at Kennedy Airport. But first he has to leave from London at King's Cross, which goes to Heathrow, and the plane from Heathrow leaves out of—"

"No, Gus, the other part," says Otto, smiling. "What does Daddy do when he gets home?"

Gus presses on with his explanation, ignoring the little detail his twin brother, Henry, whispered in his ear. Henry stands off to the side, smirking, while Gus continues with what *really* interests him: the stops on the A line from Howard Beach. I see the slightly alarmed looks on the faces of the people waiting in line. Is it the content of Gus's chatter or the fact that he is hopping up and down while he delivers it? When he's happy and excited, which is much of the time, he hops. I'm so used to it I barely notice. But in that moment I see our family the way the rest of the world sees us: the obnoxious teenager, pretending he doesn't know us; the crazy jumping bean, nattering on about the A train; the frazzled, fanny-pack-sporting mother, now part of an unappetizing visual of two ancients on a booty call.

Yet I want to turn to everyone in line and say, "You should all be congratulating us. Several years ago, Hoppy over there would hardly have been talking at all, and whatever he said would have been incomprehensible. Sure, we have a few glitches to work out. But you're missing the point. My son is ordering ham. Score!"

You may recognize Gus as my autistic son who recently enjoyed his fifteen minutes of fame. I wrote a story for the *New York Times* called "To Siri, With Love," about his friendship with Siri, Apple's "intelligent personal assistant." It was a simple piece about how this amiable robot provides so much to my communication-impaired kid: not just information on arcane, sleep-inducing subjects (if you're not a herpetologist,

I'm guessing you're about as eager as I am to talk about red-eared slider turtles) but also lessons in etiquette, listening, and what most of us take for granted, the nuances of back-and-forth dialogue. The subject is close to my heart—it's my son, so how could it not be?—but I thought the audience for this sort of thing was limited. Maybe I'd get a few pats on the back from friends.

Instead, the story went viral. It was the most-viewed, most-emailed, most-tweeted *NYT* piece for a solid week. There were magazine, television, and radio pieces around the world. There were letters like this:

> *You may be aware that right now a huge effort is being made by Apple to make Siri available in other languages. I am Russian translator for Siri, and I can say that sometimes it is very hard to transfer Siri personality to another culture. You really helped me a lot to understand, how Siri should behavior in my language, with so great examples of what people are really expecting from Siri to say. And your thesis about kindness of machine towards people with disabilities just had made me cry. We had a talk about your article in our team, and it was very beneficial for general translation efforts for Siri.*
>
> *So, with your help, Russian Siri would be even more kind and friendly, and supporting. I always keep in mind your son Gus, when writing the dialogs for Siri in Russian.*

This letter moved me deeply, as did the hundreds of emails and tweets and comments from both parents of children with autism and autistic people themselves (not that they always identified themselves as such, but when a guy tweets different lines from your piece over and over and over, you can figure it out). I think my favorite letter was from a man who wrote to the editor: "This author has a future as a writer."

Why did this story hit a nerve? Well, for one thing, it presented an opposing view to the current notion that technology dumbs us down and is as bad for us as Cheetos. But its popularity also, I believe, stems from its being about finding solace and companionship in an unexpected place. As we disappear into our phones, tablets, smart watches, and the next smart thing, it's all too tempting to disengage. These days, it's easy for everyone to feel a little lonely.

But here was a counter point of view. Technology can also bring us out a little and reinforce social behavior. It can be a bridge, not a wall.

I realized there was a great deal more to say about the "average" autistic kid. Narratives of autism tend to be about the extremes. Behold the eccentric genius who will one day be running NASA! (Well, someone has to get a human to another galaxy; you didn't think it would be someone neurotypical, did you?) And here is the person so impaired, he is smashing his head against the wall and finger painting with the blood. What about the vast number of people in between?

That's my son Gus.

* * *

No two people with autism are the same; its precise form or expression is different in every case. Moreover, there may be a most intricate (and potentially creative) interaction between the autistic traits and the other qualities of the individual. So, while a single glance may suffice for clinical diagnosis, if we hope to understand the autistic individual, nothing less than a total biography will do.

—Oliver Sacks, *An Anthropologist on Mars, 1995*

The most recent stats for an autism diagnosis are startling. In the 1980s, about one in two thousand American kids was diagnosed with autism. Today the number is around one in sixty-eight, according to estimates from the US Centers for Disease Control and Prevention's Autism and Developmental Disabilities Monitoring Network. Among boys, it's one out of forty-two. In some countries the numbers are lower; in some, higher (South Korea claims about 2.6 percent of their population has autism, as compared to our 1.6). It is the fastest growing developmental disability in the world, affecting 1 percent of all people. A University of California, Davis, study published in 2015 in the *Journal of Autism and Developmental Disorders* found that the total cost for caring for people with autism in the United States in 2015 was $268 billion, and this number is forecasted to rise to $461 billion by 2025. This represents more than double the combined costs of stroke and hypertension.

How is that possible? Researchers and writers want to

know what the hell is happening. Whether its increasing prevalence is because of something toxic in the water/air/ground, whether it's because people who were once diagnosed with other mental conditions are now labeled autistic, or whether it's because slightly odd people who used to be single are having more of a chance to breed, thereby creating even odder people (the Silicon Valley hypothesis), no one is certain. It might be a combination of all three.

Related conditions once categorized as autism, including Asperger's syndrome and pervasive developmental disorder, are now referred to under the umbrella of autism spectrum disorder (ASD). That's because the disabilities—and abilities—exist on a spectrum. Verbal or nonverbal, cognitively impaired or cognitively off-the-charts brilliant—these often very scattered abilities can exist in a confounding stew. Moreover, people with ASD don't develop skills in the kind of steady progression of neurotypical children; there's more of a herky-jerky quality to mental and emotional growth, meaning that they may not be able to do something for a very long time and then, one day, they just *can*.

My favorite story of this phenomenon involves a friend whose son didn't talk at all, save for a word here and there when he needed something. "Cookie." "Juice." He was five. One day on his way home from a party, some kids following him were mocking him in a sing-song voice about being unable to speak: "*Luke can't talk, Luke can't talk.*" After a few minutes he turned around and said, "Yes, I can. Go fuck yourselves."

I also love hearing stories about brilliant people in history who are now thought to be autistic, not because I think my own son is going to suddenly reveal himself to be some sort of genius, but because it's a reminder that the progress of human civilization flourishes with profound oddity and an ability to fixate on a problem. Albert Einstein used to have trouble speaking as a child, repeating sentences like an automaton instead of conversing. Isaac Newton rarely spoke, had few friends, and stuck with routines whether or not they made any sense. If, for example, he was scheduled to give a speech, it's said he'd give it whether there was anyone there to listen or not. Thomas Jefferson, according to Alexander Hamilton, couldn't make eye contact with people and couldn't stand loud noises. While he was a beautiful writer, he avoided communication with people verbally. The artists Andy Warhol and Michelangelo, the actor Dan Aykroyd, the director Tim Burton . . . the list goes on and on.

The phrase "When you've met one person with autism, you've met one person with autism" is a favorite in the ASD community. But there are three common denominators I can think of. One, every person with ASD I've ever met has some deficit in his "theory of mind." Theory of mind is the ability to understand, first, that we have wishes and desires and a way of looking at the world—i.e., self-awareness. But then, on top of that, it's knowing that other people have wishes and desires and a worldview that differs from yours. It is very hard, and sometimes impossible, for a person with autism to infer what someone else means or what he or she will do.

Several brain-imaging studies on autistic kids show a pronounced difference in blood flow in the areas of the brain that are thought to be responsible for certain kinds of story comprehension—the kind that allows us to know what the characters are feeling, and predict what they might do next. I thought about this recently when I was at an event for my son's special needs school, and this one guy—maybe eighteen, obviously smart and in some ways sophisticated—came over and hugged me. Then, a few minutes later, he hugged me again. Hugging! It felt good. Why wouldn't I enjoy it, too? Someone pulled him aside and told him that maybe five times was enough with a stranger. He nodded, waited till she was gone . . . and hugged me again.

Second, every person with ASD I've ever met loves repetition and detail in some form or another; if a subject is interesting to him, there is no such thing as "getting tired" of it. This can make people delightful or exhausting companions, depending how much you want to hear about, say, wind shears and suction vortices. (People with ASD are well represented in the meteorology community, I'm told. Also on Wikipedia. If you want to know who's constantly monitoring and updating pages on things like public transport timetables and the list of guest stars on *Sesame Street*, look no further than the ASD community.)

And the third? Objectively they are all a little out there. If Gus had been born in the early to mid-twentieth century, the pressure to institutionalize him would have been enormous. Even Dr. Spock, the man who famously told mothers in 1946, "You know more than you think you do," and urged them to follow their instincts, recommended they institutionalize "defective" children. ("It is usually recommended that this be done right after birth," he wrote. "Then the parents will not become too wrapped up in a child who will not develop very far.") And it's not as if the idea of a "final solution" to autism is a historical curiosity. A few years ago, in the Netherlands, where euthanasia is legal not only for incurable physical conditions but also for mental conditions thought incurable and unbearable, an autistic man who had all his life been unable to form friendships asked to be put to death. His wish was granted.

But the fact is: they are here, they are weird, get used to it. Neurologically divergent people are your neighbors, work colleagues, and, perhaps, your friends and family.

* * *

At fourteen, Augustus John is the size of a robust eleven-year-old, about four foot eleven and one hundred pounds. He has the dark, expressive eyes of a boy in a nineteenth-century Italian painting. He inherited my slightly beaked preoperative nose, which thankfully looks better on him than it did on me. He did not inherit my Jewfro, instead lucking out with shiny straight seal-brown hair. He is nearsighted, and his glasses are always smudged.

Gus has a twin brother, Henry. Henry is a head taller than Gus, blond, green-eyed, and fair-skinned. They don't look like they belong in the same family, let alone like twins.

Henry is neurotypical, which, at fourteen, is a synonym for "insufferable." Being a ferociously competitive person, Henry is always wrestling with this question: How do you prove your superiority over a twin brother who doesn't care in the least about winning or losing? Henry never stops trying. I recorded this conversation when they were nine:

ME: Gus lost another tooth today.
GUS: Will the tooth fairy take me to watch trains tomorrow?
ME: No, but she'll give you money.
GUS: Yay!
HENRY: How much will Gus get?
ME: Five dollars.
GUS: That's OK. I only want a dollar. And the trains.
HENRY: I have one that's going to fall out, too.
ME: So even *this* is a competition now?
HENRY: Do I get extra if I pull it out myself?
ME: No! What is wrong with you?
HENRY: [sad eyes]
ME: All right, fine.
HENRY: How much extra?
ME: Five?
HENRY: So, like ten total?
[Two minutes later, Henry returns with a bloody tooth.]
GUS: Yay! Henry did it!
HENRY: That was worse than I thought. How about fifteen?

Whatever the occasion, Gus is always his brother's biggest booster. It drives Henry nuts.

* * *

I'm not sure if there is a typical look of an autistic kid, though after years of school events I do believe there's a typical look of the mother: skin a little ashier than the average woman her age, hollows under the eyes a little more pronounced, a smile playing about her lips as the eyes dart about nervously, wondering what might happen next. She is sometimes proud, sometimes amused. She is never quite relaxed.

Gus's temperament? Well, my kid is almost definitely nicer than your kid. Sorry, but it's true. Your kid is almost definitely quicker, more ambitious, and more determined to take on the world. Yours will be running a Fortune 500 company, or a law firm; she will be ministering to people's bodies or souls or raising families or running marathons. Mine will do none of those things. Yours will probably be trying to figure out what ladders to climb. Mine will be delighted to be on any of the rungs at all, and I will be delighted for him. But if he becomes, say, a Walmart greeter, when he wishes you a nice day he will mean it with all his heart.

My kid tells me how beautiful I am every day, when by "beautiful" he means "clean." The bar is low. My kid can't throw a ball or button a shirt or use a knife or, sometimes, grasp the difference between reality and fantasy. But, oddly, he can play Beethoven on the piano so movingly he will make you cry. If he's gone somewhere once, he can find his way there again, next month and next year and possibly for the rest of his life. He believes, sometimes, that machines are his friends, and he doesn't quite understand what a human friend is. But he feels he has them, and he always wants more.

He is the average kid with autism. He may or may not have work; may or may not have independence; his definition of friendships and partners may be very different than ours . . . He is, like so many others, the adored, frustrating Question Mark. Maybe that's your child. Or maybe it's a child you know. And brood about. And love.

One

OH NO

It took seven years and $70,000 for me to get pregnant. My infertility started out as a mystery and with the passage of years became "Because you are old." Along the way I had five, maybe six miscarriages. I lost count. When I finally got pregnant and stayed pregnant, I threw up every day; John, my husband, would stand outside the bathroom door where I was retching, shouting helpfully, "You have to keep food down, you're killing the babies." I only gained seventeen pounds during a pregnancy with twins; postdelivery was the first and only time in my life I was thin. When the placentas completely crapped out, I had an emergency C-section at around thirty-three weeks. John insists the obstetrician told him, "We almost lost them." A retired opera singer, John is no stranger to melodrama, and I remember nothing like this. But Henry weighed three pounds, one ounce; Gus, three pounds, eleven ounces; and both did time in the neonatal intensive care unit. A baby-loving friend of mine who ran a parenting magazine dropped by to visit. She told me she knew immediately that Henry was extremely intelligent. She said nothing about Gus. Several months later she was diagnosed with esophageal cancer, and when I was seated at her bed in the hospital did not seem like the right time to ask her what she saw, or didn't see, in Gus. She died soon after. I loved her very much. And I still wonder.

Did I know something wasn't normal? Yes and no. I attributed all the little problems to Gus and Henry being twins and being premature. Whereas Gus was hypotonic—meaning his muscles were weak and loose—Henry had exactly the opposite problem. "Well, either he is going to be very muscular, or he will have mild cerebral palsy," said their pediatrician, comfortingly.

It turned out to be neither. But the fact that they had physical delays was the fog that obscured Gus's mental differences. Besides, what did I know? As an only child, I'd spent a total of zero hours with babies. If they had been dogs, I would have known that around two weeks they opened their eyes, and by eight months they would normally stop teething on my shoes. But they were not dogs or parakeets or hamsters or iguanas of any of the menagerie my extremely tolerant mother had allowed me to surround myself with. So their behavior was foreign to me. And in some sort of perverse resistance to the cult of babyhood happening all around me—I live in downtown Manhattan, ground zero for helicopter parenting—I refused to crack the cover of *Your Baby's First Year*. Milestone, schmilestone. Unless Gus and Henry had donned top hats and tails and started tap-dancing at six months, I wouldn't have known that there was anything unusual going on.

Then there was a moment.

Henry and Gus were about seven months old. Though Henry's head was so large and he was so top-heavy he'd keel over if he sat for too long, he was nevertheless sitting up, reaching for things, watching us—standard-issue baby stuff. One day my parents visited, and I was showing them what geniuses their grandsons were. Gus was in his high chair and had this mobile of twirling gewgaws in front of him, and the idea was that he would reach out and bat at the toys. I called it the Bat Mobile. In years to follow I could barely get him to *stop* spinning things. But now, at the age when it was appropriate, indeed expected, to twirl bright shiny objects, he stared off at some point in space, not acknowledging the toys in front of his face.

Hoping my parents wouldn't notice Gus's utter lack of interest in his surroundings, I picked up his little hands and punched the toys for him. Over and over again. Complete with words of praise for a job well done: "Good job, honey! See the squishy bug? Hit the squishy bug! Wheeeeee!" It was like that movie *Weekend at Bernie's*, where Andrew McCarthy and Jonathan Silverman march their dead boss around like a giant mustachioed puppet. My parents, being polite, loving, and a little bit clueless, oohed and aahed, and when they left I stuffed the Bat Mobile down the garbage chute.

At ten months the pediatrician suggested I have a home visit with an early intervention specialist. Gus was quickly diagnosed with sensory integration disorder, which as far as I could tell meant he didn't remove a sock puppet from his foot quickly enough. There were undoubtedly lots of tests, but that is the one I remember: a therapist came to our house and placed a little puppet on his foot. I think Gus's thought process went something like: *Didididididi, there's a dragon on my foot . . . Didididididi, look at those big eyes . . . Didididi. Fur . . . Didi. OK, time for it to come off.* He stared at it for a long time, whereas apparently the normal reaction is supposed to be *PUPPET—OFF.* Dawdling over the puppet is a sign that a child has poor tactile sensation and perception.

At the time I thought it was absurd, as were the other indications of Gus's alleged abnormality. OK, sure, at ten months he didn't put stuff in his mouth (no exploring), didn't look at strangers when they spun him in the air, had aversions to unfamiliar tastes and textures. The early intervention lady tried gently to explain. "There are people who go through their whole life unable to withstand loud noises, or find massage unpleasant, or can't stand the sensation of sand because—"

"Because it's horrible?" I interjected as I inched away from her to wash my hands for the tenth time that day. She was describing me. As a child I would scream if someone tried to put me in a sandbox; I am also a little frightened of anything that might be slimy—fish, okra, milk—and was thrilled to discover recently there is an actual word for it: "myxophobia." One Halloween, my cousin insisted I scoop out the innards of a pumpkin with her. That day still haunts me. Yet I managed to become a functioning adult.

And John. My husband and I have always lived in separate apartments because his apartment is a former music studio and therefore soundproofed; he hates loud noise. He is fastidious, too, and as I refuse to line up all my shoes in boxes and arrange my clothing according to texture, we both knew cohabiting was a nonstarter. (Our arrangement piques people's interest; I've even been asked to write a book about it. It's hard for me to imagine a shorter, more boring book. I always wanted the fairy tale of love and commitment just like everyone else; I just didn't see why sharing the same four walls was a prerequisite. There, that's it, and now I would have 79,975 more words to fill.) While our living arrangement might be unconventional, our marriage is filled with the affection and devotion and bickering of any other. John still buys me roses on the day after Valentine's Day (when the prices go down . . . did I mention he's a Scot?). Whatever the weather, the creaks in his back, or the pain in his knees, he makes his way downtown on the subway to greet Gus when he comes home from school to prepare his snack. And as a man utterly indifferent to sports, John will nevertheless sit for hours watching football with Henry, pretending he gives a damn about whether or not Newcastle wins. (Spoiler alert: they won't.)

So, much of Gus's divergence from everyday baby behavior didn't seem that strange to us. So what if he couldn't eat more than one food at a time, that if two were presented on his plate he would refuse to eat anything? Yes, it was true Gus first cried hysterically and then became catatonic when he heard certain sounds—the deep rumble of old elevators, for example. But what did that matter? When did slightly eccentric personal preferences become a pathology?

During the next few years, my husband and I made great use of our favorite word: "quirky." Gus was quirky. His slowness was a result of prematurity, as was his tiny size. I mean, if a kid is only fourteen pounds at nine months old, naturally things are going to take time. It was worrisome to see that at nine months he was barely a *golanim* (the Israeli word for babies who scoot on their bellies, a reference to the gun-toting soldiers in the Golan Heights war who moved by wriggling on the ground). Gus would eventually hit milestones but barely in the window where one wouldn't have a full-on panic attack. So he walked—at eighteen months. He used the potty—at three and a half years. It wasn't that he was lazy about using it or didn't understand what it was for. He did. He would just shriek when he was led there. It was so bad, and we were so perplexed, that Henry joined the fray and would drag Gus to the potty himself. Then, when that maneuver failed, Henry would use the pot and tell us the leavings belonged to Gus. When Gus got a few words, he managed to convey this: the sound of a flushing toilet was an elephant waiting inside the potty to grab him and pull him in. So after a great deal of pointing and shouting "LOOK, NO ELEPHANT!" Gus used the potty and never had an accident again.

But the words. It's not that he didn't have them. He talked late, but he did have some words by two years and continued to gain vocabulary. The problem was *how* he talked—that is, not to us.

From an email to a friend when Gus was about eighteen months old:

Gus doesn't talk yet, but it's like having a mynah bird around. He doesn't imitate humans, but he imitates other sounds. He heard a siren go by tonight, and did a pretty good impression of one. He does the microwave "bing", the refrigerator beep when the door is open. He's more interested in imitating the machines around him than the humans. But I guess it's good he's a city child. Soon he'll be doing car alarms, cars backfiring, buses emitting exhaust, drive-by shootings.

Ha-ha-ha, my child is not interested in humans!

In retrospect, it seems grotesque that I was glibly describing a quirk that was a big flashing neon sign for a more serious issue.

Gus's low muscle tone included his tongue, so he was very difficult to understand. But if he had been having actual conversations with us, it might not have been so distressing. Instead, he would greet me in the morning with a stream of words, directed perhaps at the closet or perhaps to my feet. And the words wouldn't necessarily have anything to do with what was going on. For several years, until about the age of five, he would speak in monologues. They might involve jaguars or giraffes, or simply the letter *K*, because these were the things he was fond of. They would be phrases picked up from toys or the TV or maybe even some other person, meaningless in themselves and yet declared with great vivacity. This went on throughout nursery school, and even after he learned to use a computer. He would indicate if he wanted something, but there was zero reciprocity. John and I would tell ourselves Gus was just fine because he could read when he was three; we just ignored the fact that he didn't understand what he was reading. (Many children with autism can decode words without comprehension. Who knew it was a Thing?) His language, too, was learned entirely by rote. Forget the sock puppet—if you really want to know if your child has autism, see how much he likes the announcements on public broadcasting. Gus's first real sentence, apropos of nothing, was "Major funding for *Bill Nye the Science Guy* provided by the National Science Foundation, the Corporation for Public Broadcasting, and viewers like you." Only it sounded more like "Ajorfudgforbillnyessss-guy" because his tongue didn't work.

Before autism was considered a condition unto itself, it was thought to be a form of childhood schizophrenia, and it's easy to see why: for years the relationship between reality and verbal expression for Gus was tenuous at best, and sometimes nonexistent. On the one hand Gus had many words for things, and seemed to know what those words meant, even if we didn't. But the idea of repeating what I said, practicing the language, as kids typically do? No. In fact, it became very apparent that as much as Gus loved and still loves repetition in most arenas, no amount of repetition could make him do what I was doing.

There may be a good reason for that.

The old saying "monkey see, monkey do" comes from an interesting source: monkeys. In the early 1990s, it was confirmed by scientists studying monkey behavior that when the monkeys saw the scientists eating, they signaled that they wanted food, even though they'd eaten recently. Furthermore, the parts of the monkeys' brains that signaled hunger were lighting up. (The scientists knew this because the poor monkeys had electrodes implanted in their heads. This is the kind of research that's harder to pull off on humans.) The monkeys' observation of humans eating (monkey see) triggered the same part of their brains as their own eating (monkey do). This phenomenon led researchers to discover that there are unique neurons in the frontal and premotor cortex called "mirror neurons" that help us learn behavior through copying. Mirror neurons may also make us suggestible to the behavior of others even when we don't want to imitate them. Example: Your friend just yawned. Now, try *not* to yawn. See?

In 2005, researchers at the University of California, San Diego, noted in the electroencephalograms (EEGS) of ten subjects with autism that the mirror neuron system didn't "mirror" at all. Their mirror neurons responded only to what they themselves did, and not to what others did. The implications of a dodgy mirror neuron system are profound, since the mirror neurons are involved not only in Simon Says kind of actions (a game that mystified Gus utterly) but also in all manner of learning—everything from holding a spoon to reciprocal conversation to understanding the actions and emotions of other people.

How do you learn if you don't copy? Well, in Gus's case, eventually you do, but it might be on the thousandth repetition instead of the third or fourth. To this day Gus can't brush his teeth properly without verbal prompts. No matter how many times I show him how to do it, he is mystified. As is often the case, TV and movies have come to the rescue. "FANGS," I shout, and thanks to the Count on *Sesame Street* he knows to bare his canines and brush them. Tom and Jerry help out with my other command—"BULLDOG"—and he will thrust out his jaw and brush the bottoms. But when I demonstrate by brushing *my* teeth, there is always a fifty-fifty chance he'll brush his face.

* * *

"Well, at least he's not autistic. Right?"

I cringe now when I think how often I forced all those well-meaning people—therapists, teachers, counselors, friends, babysitters, family members—to sympathetically grin through the required answer. It was the mental health equivalent of "Does my ass look fat in these jeans?" Because you know what? If you have to ask, your ass definitely looks fat in those jeans.

Everyone—*everyone*—said "No." The nursery school teacher at the fancy Manhattan preschool he got kicked out of at age four. The principal at the mainstream public school he went to for kindergarten at age five—which he flunked, even with a full-time aide. (Gus was so tiny and inattentive, one of the kids asked the teacher, "Why is there a baby in our class?") Even when we realized he would have to repeat kindergarten, and he entered a for-profit private school for the learning disabled, we were not told he had autism. The diagnosis was "nonverbal learning disability," meaning that he couldn't understand communication that wasn't verbal. Since no one had mentioned the A-word, I was all "Hey, that's not that bad!" What *was* bad was that he got chucked out of that school, too, because I didn't want to (A) medicate him at age six and (B) pay several thousand additional dollars a month (that I didn't have) for an individual teacher to be with him at all times so he wouldn't wander out of class.

Henry, never missing an opportunity to complain, wondered why he had to go to "hard" public school while his brother got all the attention at the easy fancy private school. The public school was excellent. The private school was mediocre and its director, supercilious. Gus was asked to leave.

After every debacle, I always had an excuse: Well, of course he got kicked out of preschool at four! He was deeply attached to a little girl with separation anxiety, and when she got upset he would go into a corner and refuse to interact with anyone. He was just extra sensitive! (Actually this was true, but most sensitive kids can still function. Gus couldn't.) And of course he got kicked out of the school for learning disabled kids. He wasn't on drugs like they were. (I am not at all antidrug. I was just antidrug for inattention in a child barely out of nursery school.)

Gus was six when finally a kindly neuropsychologist told

us Gus was "on the spectrum." I don't remember much about that day. I do remember that John—gruff, stalwart, very British—climbed into bed with Gus that night and sobbed.

* * *

In the ensuing months there were many tears for me, too, particularly around neuropsychological testing and schools. Neuropsych tests measure your child's overall cognitive ability, as well as his areas of strength and weaknesses.

When I tell friends I refused to look at the results, they are often shocked. They can't quite understand the paralyzing fear that comes with some kinds of knowledge. This is the only thing I can compare it to: When I was a child I had a pet boa constrictor named Julius Squeezer. The downside of Julius: he ate live mice. Every week I would go to the pet store and bring home a fat mouse in a Chinese-food container. I would steel myself as I dumped the little guy into Julius's terrarium. Sometimes the mouse tried to claw its way back into the box. It would evacuate its bowels in fear. When it landed in the cage, the mouse and Julius looked at each other and were very still. And then.

Me facing facts is like that mouse facing Julius.

While Gus's diagnosis was devastating, it did point to a general direction for his education. And the first thing we had to do was get our "special" six-year-old into the right school. The neuropsych tests we went through were mandatory for placing your child in an "appropriate" setting in the New York City school system. Now we had a better idea what that would be for Gus. There are public programs and private programs. In public programs kids with a variety of disabilities tend to be lumped together: medical, emotional, and cognitive issues in one heady brew. And while I loved public school for Henry, I was horrified by the thought of Gus—sweet, guileless, utterly unable to stand up for himself—in any public school setting with kids with an unpredictable mix of medical, emotional, and cognitive issues. The Byzantine process for getting your child into many private special ed schools involves suing the city for funding for an "appropriate" education. You have to prove that the Department of Education does not have the resources to properly educate your child. The process is complicated, but it's *there*, and for that I'm grateful to our city. Otherwise the choices were to find a way to come up with $62,000 for the appropriate school—"appropriate" being, among other things, what allows a parent to sleep at night—or put Gus into an inappropriate, potentially wildly inappropriate, school.

You know things are bad when your attorney is hugging you.

"That's great, you just cry like that when you're in the meeting tomorrow with the Department of Education," said Regina Skyer, one of a handful of attorneys specializing in suing the DOE in New York City. I love Regina. For one thing, she is whip smart, and for another, she is extremely chic, the only American woman of my acquaintance who really knows her way around a scarf. But it is her job to paint your child's situation as so dire that only the school you've decided on, whatever that is, can possibly accommodate him. Regina came to the DOE review meeting with me. She kept passing me notes in a kind of shorthand during the meeting, giving me talking points. My favorite, in block letters: "CHILD WILL BE IN JAIL WITHOUT APPROP ED." I think I was supposed to say this out loud, but it's hard to tell a roomful of strangers that you think your six-year-old is prison-bound without their help.

Regina has been wonderful, and was instrumental in getting Gus into LearningSpring, an elementary and middle school especially for ASD kids. Still, to this day, I only have to walk into her office for the tears to start. At my most recent appointment, looking at high schools for a kid who the calendar said was twelve but who looked like he was nine and acted seven, she wrapped her arms around me and said, "You know, not everyone can be the first violin. There are many positions in the orchestra!"

Yeah, I thought, *but will my guy even be able to hold up that little triangle and go* "Ting"*? Because if he can, I'll be very, very happy.*

* * *

After Gus's diagnosis and for the first few years, I was sure of many things. Starting with: there would be no real friends for my little boy. As long as he had my protection, or his brother's, he wouldn't be mistreated . . . but what if something happened to us? The milestones of a life well lived—

parties, dates, first job, first love—would be foreign to him. He would forever be the one who missed the joke.

Discovering your child is on the spectrum is like being a regular dude in *Men in Black*, blithely unaware that half of your fellow citizens are from another planet. Before the kid, what was an autistic person? After the kid—it's like, they are everywhere, but not everyone sees them. But once I could see, there were nights of pain. Not for him or myself, exactly. More like collateral anguish. The children in my past. If I had only known then.

I remembered passing a little girl wearing a bright red wool coat with sleeves and neck lined with rabbit fur. Alexandra Montenegro. Your wealthy parents dressed you so well, perhaps hoping that with the rabbit coat, the poufy velvet dresses, the white stockings, and patent leather Mary Janes you would blend in. You did not blend in. I hear your screams of frustration as the girls in the playground stripped you of that coat you were so proud of. You crouched on the concrete pavement, hands over your ears, screaming and slamming your fists on the ground as they danced around you, keeping the coat out of reach and imitating your garbled voice. The pebbles mashed into your white stockings; the scrapes on your arms began to bleed. Where was our teacher? (Only later, as teenagers, did we find out she was having it off with the principal, the Dickensian Mr. Snodstock, during recess.) This was one of those dopey private schools where a check bought admission. But, Alexandra, why did your parents, so hopeful and clueless, insist on sending you to a school where you'd be known as the retard? When the tormenting began, I would walk to the far side of the playground and pretend to study the weeds that sprouted through the cracks in the pavement. I never took part. But I did nothing to stop it, nothing at all.

Then public high school. Timmy Stavros. Jesus, Timmy, what were your parents thinking? They let you out of the house unwashed, smelling like a sewer, pants so tight you could always see the bulge, black hair slick with grease, skin so ravaged it was more like pimples that had a little surrounding skin than vice versa. You must have been going to classes, but it seemed to me that you just existed outside the school, on the perimeter, wandering and circling, like a junkyard dog. Guys were always sending you on pointless errands, just to see you scurry off obediently; you must have spent half your life looking for the janitor to report nonexistent problems in the school bathroom.

And the girls? Most just laughed; I heard one ask you for a date, then turn to her friends, cackling hysterically, and walk away. You were hormones on legs, staring hungrily, talking to no one. I was scared of you, but I vowed to myself this wouldn't be like Alexandra. So I'd say hello. That's it, just hello. I was doing it for myself, not you, and I was such a nonentity I thought you wouldn't even notice. You noticed. You'd wait for me outside classes. Your mouth would open a little, strands of saliva lingering for a second; you might bark four or five words that

I couldn't understand. Then you'd bolt in the other direction, books clutched to your chest. You ran like a cartoon character, body bent at the waist, legs spinning, kind of like the Road Runner with acne and a perpetual boner.

What was it like when you went home to your parents at night? Did you tell them you had a nice day?

Some people Google-stalk their old boyfriends. I Google-stalk Alexandra and Timmy. Timmy still lives in his parents' home in the suburbs. Alexandra seems to have disappeared. Alexandra, I want so badly to apologize. I hope you see this.

Two

WHY?

Here's the game I played with myself when I was pregnant: if something was wrong with my child, what abnormality would I be able to tolerate, and what was beyond the pale? (As you can see, I don't rate an A-plus on the Basic-Human-Decency Report Card.) Generally, a physical problem would be OK. If my baby lacked a body part, if he was too small or too big or had one eye in the middle of his forehead, I would cope. There was surgery; there was improvement. But when I thought about any kind of mental deficiency, I was lost. Not intellectual? What would be the point? There is no real life without a life of the mind.

Then I had Gus.

Like every other parent of a kid with a disability, I've given plenty of thought, usually at four a.m., to why Gus has autism. Add to that the Internet, which gives me plenty of possible reasons. Like these:

Because my husband is old. John was sixty-nine when Gus and Henry were conceived. We all know about the problems of old eggs, but old sperm—or, more accurately, new sperm manufactured by old guys—isn't doing us any favors, either. A May 2016 report in the *American Journal of Stem Cells* found that children of men over forty are almost six times as likely to develop autism as those of men under thirty, as well as being at higher risk for Down syndrome and heart defects. The increased risk is thought to involve a buildup of gene mutations in the sperm of older fathers.

Wait. Above forty is considered "older"? Hey, how about almost seventy? How does that work out?

Because I was old. I was forty when Henry and Gus were born. There is increasing evidence that older mothers are also more likely to have children with ASD, because of not only the chromosomal changes in older eggs but also some kind of environmental aging changes in the uterus. Great. As if having saggy tits weren't punishment enough.

Because I was fat. OK, not fat. But certainly not svelte. And large epidemiological studies have shown that maternal obesity and gestational diabetes have been found to increase the rate of autism in children.

Because I had twins via IVF. Apparently it's not so much the in vitro fertilization, per se, since there is no increased risk of autism in a single birth. But when twins or triplets result from IVF, there is a higher incidence of one of the children having autism, leading researchers to believe, again, that uterine environment as well as genetics plays a role.

Because I was a vitamin junkie. Once I got pregnant, I was a little selective about the nutritional rules I obeyed. I nobly gave up smoking and sushi, having never tried either to begin with. I stopped drinking, because I could, and also because I was throwing up all the time and almost never even wanted it (though on the few occasions I did, I would just read something reassuring and French).

But when it came to rules I could follow that didn't involve sacrifice, I was all for them, which is how I started downing prenatal vitamins like Skittles. Essential for staving off birth defects! I was on it! But wait, I recently discovered that according to research conducted by the Bloomberg School of Public Health at Johns Hopkins, high levels of folic acid at birth are associated with a rate of autism double the average. And very high rates of B12, also in prenatal vitamins, triple the chances the offspring will develop ASD. And what if you have very high rates of both? Good times. The chance that you will have a child on the spectrum, according to this 2016 study, increases *17.6 times.*

Because of 9/11. September 11, 2001, was the date the World Trade Towers fell. September 25, 2001, was the day Henry and Gus were born. My home is about a half mile from where the Towers stood. Although I was hospitalized farther uptown for the last few days of my pregnancy, I lived for almost two weeks with the rancid metallic stench of the detritus of those fallen buildings. A 2014 study by researchers at the Harvard School of Public Health found a strong link between autism and in utero exposure to air pollution: the risk of autism was doubled among children of women exposed to high levels of particulate air pollution during pregnancy. Sooo . . . I was exposed at the end of the pregnancy, not the beginning, when one would think structures of the brain are forming. But who knows.

Because there is something weird about John. As long as I've known my husband, he's been an opera singer. But in his twenties, before he realized he could make money by opening his mouth and belting it out, he was an electrical engineer. According to a series of British studies, the children of engineers are about twice as likely to have children with autism, and even the grandchildren of engineers are more commonly affected. Autism has clustered around major engineering centers, like Silicon Valley; Austin, Texas; and Boston's Route 128 Technology Corridor.

Why would engineers produce more autistic children? Well, it's not that they are engineers, per se; it is that they are "systematizers"—the kind of people who see the world in predictable, repeatable, law-governed patterns. The other pole in human behavior are "empathizers," those who see the world's events as more random, and more governed by the vagaries of human emotion. So whereas an empathizer might come to a crime scene and ask first, "What's the killer's relationship to the dead person?," a systematizer would want to solve the crime based on blood-spatter patterns and bullet trajectories.

Everyone's outlook on life falls somewhere on a continuum, of course. But engineers and their ilk are at the farther reaches of the systematizer side of the spectrum—and so are autistic people. Emotional motivations might be a little more mysterious to them, as anyone who's ever been involved with an engineer (or computer programmer, or basically any guy who works in a lab) can attest. I remember one date in graduate school with a boy who later became one of the leading experts in the country on virtual reality. I produced these body paints I thought we'd try out. Did I mention I was nineteen? Anyway, he was very interested—in the chemical composition of the paints. How long would they take to dry when applied? Were they actually edible or merely nontoxic? And would they retain their pigment once painted on a reactive canvas like skin? You can imagine how well that date went.

My point is, engineers are not known for their mad

dating skillz, and in prior centuries they might not have found mates. But Simon Baron-Cohen, director of the Autism Research Centre at the University of Cambridge, theorizes that part of the reason for the rise in autism levels is that people who are socially inept are more likely than in days gone by to find partners and breed (thanks, Tinder)—often with women who are similarly technologically gifted but socially awkward. Thus, the rise in the number of children who, in Baron-Cohen's parlance, are systematizers—sometimes leading to extraordinary talents, sometimes extraordinary disability. And sometimes both.

John has a son, Karl, from a previous marriage. Karl is in his sixties, considerably older than I am. He is a wonderful painter, local historian, and old-bottle collector who doesn't have a phone, let alone a computer. He knows everyone in his small village in Northern England, but has no close friends. When John goes to England twice a year, there's no need to make arrangements to get together with his son. John just turns up at the pub Karl has been going to every Saturday night for the past forty years and finds him there. Karl lived with his mother, John's former wife, until the day she died. He never married, never had a girlfriend or boyfriend. He seems content.

"In England during that time, one never considered *that*," John said to me recently.

I wonder about John. I've asked him repeatedly if he

might take a short psychology test called the Autism Spectrum Quotient. He is always too busy.

Because there is something weird about me. People with ASD often have a sensory system that doesn't function properly, either underreacting or overreacting to stimuli in the environment. Exactly how the brain malfunctions is a little unclear. It has something to do with the interaction between the sensory receptors of the peripheral nervous system (the body minus the brain and spinal cord) and the central nervous system itself. But sometimes one person can be both overreactive in some circumstances and underreactive in others. Gus is overly sensitive to heat: he wants all his food at room temperature, and washes in cool water that most of us would find too cold. On the other hand, he is underreactive in his sense of space: he still knocks into people on the street, and would talk to me about two inches from my face if I weren't constantly grabbing his shoulders and pushing him back.

To a lesser but still marked degree, I have the same problems. My entire morning can be ruined if I touch something sticky. I have a sizable collection of unused gift cards for massages from well-meaning friends who don't know my rule: if I'm not having sex with you, I don't want to be touched. And when I'm sick I often have synesthesia, a scrambling of the senses where sights, sounds, and tastes can become peculiarly jumbled. So when, for example, I was pregnant and nauseated I had to keep my eyes closed and lie still every morning. I had red bedroom walls, and if I looked at them, I could smell and taste rotting meat.

Because I was a mess. Until this year there were studies showing a weak link between taking antidepressants and an increase in children with ASD. Now, it seems, Massachusetts General Hospital has debunked that link. But there is still a correlation between depression and anxiety in mothers and ASD. I did not take antidepressants. But maybe I should have, since I was in a perpetual state of dread. Money worries; constant nausea; a sense that I did not have the patience to be a mother to one, let alone two; an aging husband who did not buy my argument that "If Larry King could do it in *his* dotage, so can you"—all of these added up to a deeply unhappy pregnancy. Was overseepage of cortisol, the stress hormone, wreaking havoc with Gus's teeny brain? And even now, sometimes I see a direct correlation between this time in my life and his own irrational fears. Every time Gus hides in the closet during a thunderstorm, I think, *If only I had taken pregnancy yoga.*

So, to recap: older father + reproductive technology + twins = trifecta of bad juju. These are three main risk factors for autism. I had them all.

But there was yet one other possibility:

Because I am a bitch. A few days ago Henry said to me, "I think you were born to be a mother."

Ha. Ha. A-hahahahahahahahaha.

After Henry and Gus were born by emergency C-section, I did not see them for twenty-four hours. Not because I wasn't allowed to. I just didn't feel like moving. I had already told the nursing staff at NYU I wasn't going to breast-feed. Instead, I alternated between sleeping and pushing the button on the morphine drip like a lab rat. Friends and family saw my kids before I did.

In the months that followed, I shelled out money I couldn't really afford to keep a baby nurse because I wanted to keep working and I knew that sleep deprivation meant no work. That was half the reason. The other half was that I was actually rather frightened of babies, and having two that were the size of roast chickens and looked a little like that baby in *Eraserhead* didn't help their cause. John was also not a source of solace. In retrospect it's understandable that a man his age who already had a grown son—a son older than his current wife—would not be that engaged by infants. He never changed a diaper, and was given to saying things like, "Children destroy your soul." Eventually he turned out to be a more patient, nurturing person than I am. But at that point I spent most of my spare time either crying or looking at men on Match.com—many of whom, I told myself encouragingly, weren't currently aware they wanted a forty-one-year-old woman with twins, but would know the moment they laid eyes on me.

Love for Henry and Gus came ineluctably but gradually. Those first six months were brutal.

No matter that the idea is debunked, as it has been for so many years, every mother of a child with autism occasionally flashes on the term "refrigerator mother," popularized by famed psychologist Bruno Bettelheim. (Actually he was a lumber merchant who then studied philosophy at university in Austria. He then reinvented himself as a child psychologist after moving to the United States.)

Autism was first introduced to the popular culture in a *Time* magazine article published in 1948. Titled "Frosted Children," it reported on the work of Leo Kanner, the psychiatrist who in 1943 had identified the condition of "early infantile autism." Kanner always believed that autism was innate. But he also speculated, perhaps a little unwisely, that the parents of these "Diaper Age Schizoids," as the article calls them, were unusually cold and did not really exist among the less well-educated classes. Kanner later came to believe that the parents themselves had traces of the autistic behavior that found full flower in their children—in other words, the parents had some autistic traits and passed them on.

But that's not how Bettelheim, who directed the University of Chicago's Sonia Shankman Orthogenic School for emotionally disturbed children, saw it. Bettelheim believed the parents' coldness *caused* autism. The part of his argument that haunted me? This, from Bettelheim's 1960s bestseller, *The Empty Fortress*: "The precipitating factor in infantile autism is the parent's wish that his child did not exist. Infants, if totally deserted by humans before they have developed enough to shift for themselves, will die. And if their physical care is enough for survival but they are deserted emotionally, or are pushed beyond the capacity to cope, they will become autistic."

Now, why would I spend years shooting myself up with fertility drugs and copulating on command if I didn't want children? And yet in dark moments I wondered. Were all those miscarriages the mind guiding the body to do what I really wanted?

When I needed still more ways to blame myself, I would read about the genes involved in brain development, more specifically about the two hundred plus mutations that are found in children with ASD. Are some important? One? None? All? Then, I came across the studies on mice that show that if they contract certain viruses during pregnancy, their babies will develop autism-like symptoms. Did I have a bug that set off a chain reaction of brain damage? Who knows. I was too busy throwing up.

I still have days when I work myself into a frenzy about all the ways I could have caused Gus's autism. But there is that theoretical awfulness, and then there is the precious small person. I look at Gus the person, not Gus the mental condition, and I calm the hell down.

Three

AGAIN AGAIN AGAIN

ME: Have I mentioned that you're handsome?

GUS: Yes.

ME: Have I mentioned that I love you?

GUS: Yes. And have I mentioned that you're bad?

ME: *MEH.*

GUS: [chortles]

GUS: What's the high point of your day?

ME: Putting you to bed.

GUS: And low points?

ME: No lows. And you?

GUS: The high point of my day is when you put me to bed. No lows.

ME: Now, I have just one question for you . . .

GUS: [Shining eyes. Wait . . . Wait for it . . .]

ME: Are you my sweetheart?

GUS: *YES!* Yes, I am your sweetheart.

"Mom? Seriously? You're *still* doing that?" Henry says, incredulous, when he overhears our nighttime catechism. At fourteen, Henry rarely leaves his room at night except to walk the eight feet to the refrigerator for snacks, which explains how I've been hiding it from him for a couple of years. OK, maybe ten. Henry thinks that logic is the answer to everything.

"First of all, it's not true," Henry begins. "He's not that handsome, frankly, and if the high point of your day is putting him to bed, you need to get a life. Second, it doesn't even make sense. When he says, 'Did I mention that you're bad?' and you say 'MEH,' did it ever occur to you to challenge that assumption?"

"It makes him laugh," I say, a little defensively.

"I know . . . but it made him laugh when we were *five*. WHY ARE YOU STILL DOING IT?"

Why indeed. Why do I let him eat the same food, wear the same pajamas, watch the same videos, give me the same weather report at dinner, and rescue his stuffed animals from the garbage when I try to throw them out? ("They like to sleep with me.") These eccentricities are minor. But others are not, because they reveal an inability, or unwillingness, to learn from experience.

When I got a note from Gus's school that they were concerned about his hygiene, I was mortified. With the amount of reminding and nagging I've devoted to making sure both my sons brush, shower, and wash their hands, it's miraculous that I have not induced some form of OCD. I tell Gus, again and again, that if you are autistic it is particularly important that you practice "handsomeness" (as he calls it), so that even if you are different from other people, they are not going to be put off by you. I believe this, and I say it every day of his life. But there comes a certain point when a mother does not want to be washing her son's hair for him, and he doesn't want her to, either. The problem is that Gus cannot get the idea of washing out the shampoo. He cannot, will not, put his head completely into a shower stream and tilt his head back; he seems to think he will suck in water and suffocate. So if I don't physically force him, the soap builds up and his hair looks like it belongs on Johnny Rotten. It's not a big deal, and yet it's a big deal. I come close to tears over the soap in his hair. Sure, I could solve the problem by giving him a buzz cut. But when I've done this in the past, his opera-singer father, whose ideas about men's hair were unduly influenced by his many performances in *Samson et Dalila*, just looks too sad.

* * *

All kids enjoy, and need, a certain amount of routine in their lives. But most are also wired for variety. People with autism are wired for predictability. Sameness is Gus's jam.

I have to be careful about the sounds I make around him because if he loves one he will demand I make it for the next infinity years. And usually the word has to be sung, not spoken. So he will only brush his teeth if I sing "Mint!" and only wash his hands if I say, in a querying manner, "Palms?" (For years teachers called him the Boy with No Hands because he couldn't let anything touch his palms. Only fingertips ever saw water. Thus my insistence on washing palms.) And food? He has had the same plate of apples, bananas, and Cheerios every morning since he could eat solid food, and every night, the same rice pudding. (They have to be the same kind of apple, too. When we'd go apple picking in the fall, he would gather them happily but not want to try them, because he knew they were not Fujis.) Every Friday night without fail there are chicken fingers and fries from the local Greek diner. Mashed potato is the only potato, and he devours an avocado a day. No vegetables, no rice, no bread, no pasta. He likes exactly the same foods and greets them with the same relish, no matter how many times they are served.

Being such a cheerful fellow, he does not have a complete nervous breakdown like many young adults on the spectrum when his routines are broken. If we walk a slightly different route to school, he merely trembles but doesn't throw himself on the ground. If a train stops unexpectedly in the tunnel, tears might stream silently down his face, but he won't get angry. And if, say, a subway has been rerouted and the E train appears where the B train was supposed to be, I can always jolly him out of his panic by explaining that it was a "magic" train. Indeed, saying something or someone is "magic" often softens the blow. Which is why the substitute teacher at school—the unexpected person—is practically a sorcerer as far as Gus is concerned.

Still, the anxiety for the unknown is ever-present. No amount of reasoning has ever been able to make him stop crying when he sees on his beloved Weather.com that there might be a thunderstorm. "I know it won't hurt me," he says as he grabs his bedclothes and drags them into his closet, where he will spend the night. "I just don't like the noise."

And while that's true, it's not even that. It's the uncertainty. I was very pleased with myself when I bought a lightning-tracking device that tells you when there will be a strike within a certain range. That sounded like such a good idea ("See, honey, you'll always know!") until I discovered that the range was something like twenty-five miles. This may be useful when you're in the middle of a farm in Kansas, but when you're in New York City a lightning strike twenty-five miles away means nothing, except that now your terrified child is constantly on the lookout for lightning and thunder he would never even notice, save for this stupid device you just bought.

I had to disappear the lightning detector, but not before there were several miserable nights cowering in the closet. Today statistics mean something to him, so he is cheerful when AccuWeather reports there is a 20 percent chance, or less, of thunder and lightning. Still, when Henry wants to get Gus running to the computer in fear, all he has to say is, "Hey, Gus, isn't there a 75 percent chance of thunder tonight? *Why don't you look it up?*"

* * *

"So what if he likes routine? I don't understand why you get so upset," John says as he and Gus finish up the same conversation about changes in the train schedule they have had every night since Gus could talk. Then John hears the too-loud music playing upstairs, shouts in the general direction of the neighbors despite the fact that they can't hear him, and resumes eating dinner, one of the five or six acceptable entrées he's been eating repeatedly for the past eighty-three years.

The love of sameness is what first drew me to my husband. After a lifetime of tumultuous relationships where, safe to say, fidelity didn't play a great role, John offered sweetness and constancy. Because he is a person of such stolid habits, I figured he'd never get bored with me. I mean, so what if he likes the same completely bland overcooked food, cooked in the same way, every day of his life? It makes him happy. He seeks out the same kind of wide-lapel suits he wore in the 1950s, refuses to go to restaurants "because you never know what they sneak into the food," never changes his opinion on something once his mind is made up, refuses to learn to use a computer, and scorns social media—"because why do you need the world knowing your business?"

He loves classical music and rereads the same great books every year. I admired his intellect, even as I wondered at his lack of curiosity about anything new. And sexually . . . did I mention that part about never getting bored? All I had to do was move to the left or kiss to the right and I was a wild woman. This assuaged my deepest insecurities. The fact that he didn't want to live with me because there was too much chaos in my life suited me fine. If someone loved you but got upset when the pillows on the bed weren't lined up properly or his special mug wasn't in the section of the cabinet where he expected it to be, you, too, might think it unwise to live together.

So Gus's love of routine didn't seem so odd to me. At first.

I'm not sure when I noticed his love of repetition far exceeded anything other toddlers loved. Maybe it was the Mozart cube he carried around with him until he was seven, playing the same tunes, in the same order, over and over again. Maybe it was the fact that he, like his father, could not leave a room without straightening it out and closing all the drawers. (Some compulsions are useful.) We all understand the thrill of the familiar when it comes to certain things—music, most obviously, and poetry, and lines from favorite books. For me, it's the first line of my favorite twentieth-century novel: "Lolita, light of my life, fire of my loins. My sin, my soul. Lo-lee-ta: the tip of the tongue taking a trip of three steps down the palate to tap, at three, on the teeth." Seriously, that never gets old.

But how can the same be said of a video of wooden escalators at Macy's? This video on YouTube lasts fifteen minutes. Some guy filmed himself going up every floor. The last time I looked, the video has 430,469 views. I'm sure that (1) every person who has watched it in its entirety is on the spectrum, and (2) 300,000 of those views belong to Gus.

Gus has a word for the unfamiliar, and that word is "no." When I told him that I wanted to change the curtains and bedspread in his room, or maybe just take down the finger paints from second grade, he looked at me plaintively and said what he says about pretty much everything: "Mommy, I just like the old things best."

Every parent or caretaker of an autistic person has his or

her story of Same. My friend Michele remembers getting her brother to bring her twins to visit her in the hospital when she'd had her third baby. One of the twins is on the spectrum. Every day for three days, they took the subway. "My brother had to walk through the cars asking people to move because my son Jack would sit only on yellow subway seats, never on orange. People would point out an empty seat and my brother would be all, 'No can do. Gotta be yellow.' He didn't even explain why. He was just like, 'Nope, he *needs* the yellow seat.'"

In April 2016, an article in Spectrum, the online autism-research-news site, explains why up to 84 percent of children with autism have high levels of anxiety, and up to 70 percent have some sort of sensory sensitivity: they are lousy at predicting the future. They tend to miss the cues. But it's not like they're golden retrievers, living forever in the present. They know perfectly well that there *is* a future. So combine these two concepts—knowledge that the future is coming and being horrible at figuring out what it might be—and you can see how knowing your breakfast will *always* be apples, bananas, and Cheerios might be extremely soothing.

Repetition also makes us feel competent, a phenomenon not reserved for the autistic. My friend David Kleeman, one of the leading children's media experts in the country, explained to me that repetition is purposely worked into children's programming. "Nickelodeon would put the same episode of *Blue's Clues* on five days in a row," he told me. "Day one, it's new to the kid. Day two, it's familiar. Days three to five, the kid anticipates the interactivity and feels smart." When Gus was little, he failed abysmally at those sequencing tests where you are showed a bunch of cards and asked to put them in an order that makes a logical story. He's not much better at them now, because the ability to infer is damaged. So it must be an awfully good feeling for Gus to know what happens next, which is part of the reason that he still watches *Sesame Street* videos. He will speak all the lines or play different parts; he seems to have an affinity for Ernie. If you think this is alarming and frustrating to see in a teenager, you are correct, but everything is relative. He has finally lost his taste for *Teletubbies*. Or has gained enough guile to hide it from me. Either way . . . *yay!*

And then there is the possibility that what is excruciating sameness to me holds thrilling shades of difference for him. For many years Gus would stop everything he was doing to watch the ambulances going by our window, and would usually know which hospital the ambulance was coming from. I figured he just had good eyesight and could see quickly as they zoomed by. But then, I discovered he was myopic, but it didn't matter—he knew in the dark. It turns out that the sirens have slightly different pitches—and since he watched ambulances on YouTube, he associated each pitch with the particular hospital and knew which ones were passing by. This is just one of my many boasts when I get together with other parents of autistic kids and we play our favorite game, "Why Isn't This a Marketable Skill?" Oh, sure, other mothers can talk about their child's grades and the number of parties they're invited to, but can your child tell which person in a family slept on a particular pillow by the smell? Or can your little genius tell when food is going to go bad before anyone else? My friend Andrea's son can remember everything that happened on a particular date—"but only for himself. It can be as large an event as a trip to an amusement park or as small as a trip to the store. He can go back years, and tell everything about the day in chronological order. If only this memory applied to academics."

If only.

* * *

Oh my God, though, the boredom. Before Gus, I'd always seen ennui as a character failing: if I was bored, that was my problem because the world is too fascinating a place. I now know when I began to question this belief. I was looking back at an email I'd written to a friend, venting about John. It was long before we had kids. Because we never lived together, we tended to call each other frequently. John lived on the Upper West Side of Manhattan, and I lived downtown. This was the email:

It's 9 am, and John and I have already had three conversations about the subway. I live in dread of changes in the subway system, because every time the #2 temporarily goes local or the N & R aren't stopping on Prince Street, I must listen to several daily updates. The phone rings. "Good morning! Did you hear that today the uptown F is not stopping at . . ." and I want to scream, HELLO, HAVE YOU NOTICED THAT A) I BARELY EVER LEAVE MY HOUSE AND B) I DON'T GIVE A SHIT? But if I didn't listen to this, I think he'd be calling up strangers and discussing the MTA with them.

It's one thing to admit that your husband's love for unwavering routine is tiresome, but who wants to admit your own child bores you? And I don't mean just the boredom of being with a baby—what my friend Moira describes as "hanging out with a crapping, screaming roast beef." I'm talking about the boredom of being with a sentient human. Teenagers may be many things, but dull is usually not one of them. Henry is an obsessive sports fan, reluctant reader, and hater of musical theater, meaning the overlapping area in the Venn diagram of our common interests is the size of a pinhead. But at least we bond over politics and cute animal videos. Why couldn't I find something to share with Gus?

We all have different ways of coping with the boredom. For me, all it takes is a couple of cosmos, and I can feel my frustration slipping away. After thirty years living here, I still have a crush on New York City. And there is such promise right outside my door. All I have to do is slip on this little black dress, wedge myself into my highest heels. Then, soon, the throbbing of the music, the square-jawed stranger waiting for me, and the zipless . . .

"Mom, have you been hitting the sauce again?" Henry says the next morning as he looks around the kitchen. Lemon carcasses and Brussels sprouts are scattered everywhere. My head hurts.

What I actually do when I'm bored and drinking is search Epicurious.com for something I can make *that very minute* with ingredients around the house. Family, I spit on your bourgeois pizza-and-chicken-finger lives! Under the influence—which might mean two, maybe three drinks, because I'm an animal—the idea of making Brussels sprout chips seems a sound one, even if it means spending an hour in the kitchen separating the sprouts into individual leaves. Then, I guess I must have thought that the chips would need a spritz of lemon, and I found I had an entire bag, so I got out my hand juicer and went to town. Finally I must have exhausted myself, because I didn't actually make the chips but stuffed the leaves in a plastic container in the refrigerator and left the juice in a flask on the counter. Later I found I had grated the lemon peel and placed it in the freezer, neatly sealed in a Baggie.

But for a couple of blessed hours around midnight, I was not discussing weather, trains, or Disney villains. PS, Brussels sprout chips are ~~awesome~~ ~~tasty~~ ~~close to edible~~ a totally lame food.

* * *

Recently Gus and I were standing at the top of a very steep escalator, looking down on people's heads. In the past ten years I have probably spent more time watching escalators than going to movies. While there, I try to see what he sees.

There are the people fidgeting and bobbing, the unpredictable swirl of color and motion. They are juxtaposed against the predictable grid of silver marching stairs. It's the stairs he loves most of all. "Mommy, look how beautiful," Gus says, enthralled. "Look!"

I see people riding an escalator. Maybe it requires magic mushrooms to see it the way he does.

* * *

The desire for repetition and predictability is not just emotional—it's a physical need as well. Hence the autistic behavior known as stimming. "Stimming" is the shorthand term for self-stimulation. No, not that. It's a repetitive behavior that is both calming and pleasurable to people on the spectrum—rocking, flapping, spinning, or, in Gus's case, making the click-clack sound trains make. He'll do it with toy trains, but when they're not readily available he'll do it with pencils, or maybe salt and pepper shakers. This makes for good times at restaurants.

One of great debates among parents is: To (let your child) stim or not to stim? Parents ask themselves, *What else could my child be doing instead of spending hours in a seemingly meaningless activity?* I pondered this for many years, heard differing opinions from parents and teachers. Finally it occurred to me to go to some people who really might have the inside scoop on stimming: autistic adults. And while there's debate about how much stimming and when, the consensus seems to be: Let 'er rip. Because if the person *doesn't* spend at least some time in the day stimming, he can pay a steep price.

Maxfield Sparrow is an autistic trans man who describes himself as "nomadic"; he lives in his van with his cat, Mr. Kitty. (Not for the alarming reasons you might imagine. He has a sleep disorder that has been helped immensely by constant exposure to natural light and darkness.) He is also a writer whose work deserves a wide audience, particularly an audience of neurotypical people who want to understand their autistic loved ones.

On his blog, unstrangemind.com, Sparrow explains that stimming occurs both in times of stress and times of great happiness. "I have as many different ways of flapping and twisting and ruffling and fluttering my hands as I have emotions and emotional combinations that wash over and through me. My hands are like barometers of my emotional climate," he writes. Stimming is not so much a way of exhibiting control over the environment as an outlet that the stimmer can control when things aren't just so—because Sparrow's need for routine and predictability are profound. "I do . . . things like always removing the ice cube trays from the freezer in the same order, always putting the same number of ice cubes in my glass, always walking or bicycling the same route to get places, always brushing my teeth for the same number of minutes every night, and so on. These things serve my need to have a predictable, orderly world that is under my control as much as possible. The more I am able to feel a sense of control over my life, the calmer and happier I am." Everyone feels this way to some degree, of course, but to autistic people it's often to a much, much higher degree.

But here's an idea Sparrow introduced me to that I hadn't considered before. He describes himself as "alexithymic," a term that refers to having deep emotions, yet not always being aware of them or understanding what they are. Stimming behavior actually clues Sparrow in to his own emotional landscape. "Without my hand flaps, I would not be anywhere near as connected to my inner life. Without my hand flaps, I would struggle so much more every day, just trying to understand what my body and spirit were experiencing. My hands are my teachers and they educate me about my deepest self every day."

Neurotypical parents of autistic kids need to pay attention to autistic adults. Maxfield Sparrow taught me this: Your kid is doing something that looks weird yet is bothering no one? #1) that weird thing has a real function. #2) Get over it.

* * *

The passion for sameness and repetition will probably never leave Gus. In a person of great cognitive ability, our world can change because of autistic single-mindedness. Isaac Newton, rumored to be on the spectrum, did not discover gravity in between his many other hobbies; he just thought about this subject *all the time*. In a person of Gus's more modest abilities, it means that when he's hanging out in our lobby, every person who is unsure how to get somewhere by subway knows that all she has to do is ask Gus and he'll tell her the quickest route. Lots of neighbors know they don't need to check Google Maps anymore. Gus is their Google Map.

I do my best to put Gus's desire to do the same thing all the time to good use. He is my little Sherpa, happily running up and down the stairs to shut off the lights if I am too lazy. Every night, right before I go to bed—and whether I need it or not—he brings me a glass of water, with a great flourish. To get him to move off a piece of music he loves is difficult. But then isn't that the essence of practice?

Does he even understand that most people are not entranced by escalators? That he doesn't see the world the way most others do? I've tried to approach the question a few times—"Do you know you are autistic?"—and he always acts like he doesn't hear me. I want to understand what he's thinking. *Is* he thinking?

I keep trying.

* * *

I am a seeker of novelty. Most journalists are. But my own love of the New, powerful as it is, doesn't stop me from occasionally grasping what Gus feels.

I have an office in my apartment building, three floors above my home. One day I was pontificating to my officemate, Spencer, about how I don't understand why people cling to useless material things, and after listening to me gas on for a while, he said quietly, "Your parents have been dead five years. You have everything in a storage locker. You've never even gone up there to look."

That's where he was wrong. About a year after my mother died, I went up to the huge storage unit that housed a life's worth of goods. It smelled like the mold that took over my parents' house in the last years of their lives. Much of it was worthless, though I thought there could be some treasures among the dreck. I had an appraiser come up there. We were going to go through everything. I would be ruthless.

I opened the door, looked in the unit. There was one particularly ugly lamp that had stood next to my mother's bed for forty years. It was brass, and had little cherubic figures dancing around its base, accompanied, for some reason, by sheep. It would be the first thing I'd dump.

I stared at the lamp, shut the locker, paid the man for his time. It's been five years of paying storage for things I've never liked, even when I was a kid. And I'm still not ready.

Same is sometimes not a choice.

Four

I, TUNES

Gus got a card in the mail today. It was a CD, plus a letter from Laurie.

LAURIE!

Hi Gus!

It makes me so happy to know that you still enjoy my music. You are such a wonderful person and I'm so glad I know you.

I hope you never stop making and enjoying music!

Love, Laurie

She signed her CD:

Keep on singing—and shining bright!

Laurie Berkner

Now if you don't know who Laurie Berkner is, that is because you have not had a child under six in the past decade. It means you cannot recite the lyrics to "Victor Vito" in your sleep, and furthermore you do not know that Victor

Vito and Freddie Vasco (who ate a burrito, with Tabasco) are composite names of Laurie's accountants. Or that "We Are the Dinosaurs" ("Marching, marching") originated as a song when Berkner was teaching preschool, and she wanted a song that could help the kids work off a little steam. You also are unlikely to have a special Laurie-designated stuffed animal so that when you're attending a concert and she starts in with "Pig on Her Head" you are prepared with your own personal pig (or cow or, in our case, giraffe, because that's how we roll). You have not, in other words, experienced the Phish of four-year-olds.

And that's too bad, because she's actually a wonderful singer and composer. None of this "Baby Beluga" crap that'll make your ears bleed. Laurie Berkner's music is so catchy that I used to suspect people borrowed other people's kids just to go to her concerts.

Of course, if your child is over eight, he or she has probably moved on to Taylor Swift, Katy Perry, or, God help you, Miley Cyrus. Laurie is a memory for a certain enchanted period of youth. Your kid is not going to be subbing "We Are the Dinosaurs" for Mendelssohn's "Wedding March" on that special day.

At least, most kids won't. But there is fandom, and there is autistic fandom. This is the story of Laurie and Gus, and how sometimes, no matter what, there are people and places you just can't quit.

* * *

Nobody knows for sure why music means so much to so many people with autism. There are theories, though. Brain-imaging studies on those with autism show abnormal activity in areas of the brain associated with language and the processing of social information, like faces. But the parts of the brain that are receptive to music are undamaged, and may even be particularly well developed. Autism is considered a communication disorder, but before there was language there was music—or at least that is what more and more evolutionary scientists think. (Charles Darwin talked about our ancestors singing love songs to one another before they had articulate language.) Repetition, rhythm, melody, tone, duration, volume—all can reach even profoundly autistic people in a way language and visuals often can't.

At various points in his life Gus has taken music therapy, singing lessons, and then, when his voice began to change, piano lessons. Whatever the class, it was always the highlight of his week.

The musical group therapy was, to my ear, a group of kids banging on random instruments. In fact, it's much more. Alan Turry, Gus's teacher at the Nordoff-Robbins Center for Music Therapy, has seen music reach kids who never spoke or reacted to anyone. Turry believes that certain scales reach them before they are ready for more sophisticated ones. Pentatonic scales for example—used in Chinese and folkloric music—are open-ended, and don't call for resolution the way dissonant chords do. They are seductive and meet you on your own terms, he says.

"It's wrong to generalize about people on the spectrum, but I will say that music can be not only a connection, but eventually a first form of conversation, a back-and-forth that people who can't use words can take part in," Turry told me. For those like Gus who do have words but still have trouble expressing themselves, music can be a language more fluid for them than regular conversation. "Gus is so musical that in some ways he may gain more self-awareness through music than he does through conventional language," Turry adds.

Turry then illustrated his point with something I'd forgotten from years ago. When Gus was little, I could never get him to wait his turn. I chalked it up to impulsivity, part of his condition. But in fact he could wait just fine if he was asked to do so in music therapy class. When he was sitting with a bunch of kids playing percussion instruments, he learned about turn taking easily because his instrument organically came in at a particular time in a piece that they played. Ergo, music could convey an idea to him where I couldn't.

When I think about it now, this has always been true. From the time he was a baby, sound had meaning to Gus where words did not. When we talked to him, Gus generally didn't pay much attention; often he wouldn't even turn his head in our direction. But if we played music, he'd snap to. A music box—literally a plastic box with sides that you smacked and snippets of classical tunes would come out—was his constant companion for years. He would cry if I played certain songs, particularly the theme song to *Cheers*. (Hell, I could make myself cry with that one. Plaintive melody + the lyrics, "You want to go where everybody knows your name" = *Wahhhhhh.*)

As he got older, he would listen to my iPod for hours, staring at the titles as they cycled through. He didn't always listen to the songs in their entirety, which annoyed me. But clearly he absorbed them. That iPod resulted in his favorite party trick: identifying any of the hundreds of songs from just two or three notes and, often, just a fraction of *one* note.

I still have no idea how he does this. But I do know that if someone reinstates that old game show *Name That Tune*, Gus will be a millionaire. As it is now, Henry has turned this talent into a betting game with unsuspecting friends. Henry needs to get some new friends, because by now they all know not to bet against Gus, and his source of easy money has dried up.

From an early age Gus loved Mozart and Beethoven and Chopin, and he had perfect pitch, a talent that pleased his father, the singer. This gift is not even unusual. But at the same time, the pleasure of finding out he was musical at least softened the blow of his many limitations. Plus, it's kind of nice having a karaoke bar in my own home, as Gus tends to sing along with whomever he's listening to on YouTube. He sings as much as he talks.

I wanted to play up his strengths, so when Gus was seven, in addition to the music therapy classes, I enrolled him in a voice and music class with neurotypical seven- and eight-year-olds. He was kicked out after one session. He got the idea of reading music right away. But then he wandered off, put his fingers over his ears, sat in a corner, and started making train noises. I felt an overwhelming despair for him; he couldn't do the thing I knew he loved. The teacher saved the day, as teachers often do. "He can't have lessons in groups," she said, "because when he hears people singing off-key, it kills him."

Of course. I should have remembered. Now, Gus's inherent way of seeing the world means he never says a bad word about anyone—every woman is pretty in Gus's eyes, every man handsome—but there is one exception. I love to sing, know the lyrics to dozens of musicals, and am tone-deaf. As soon as Gus sees me preparing to belt out, say, "Oklahoma," he claps his hands over his ears and runs out of the room, screaming, *"NO NO NO NO NO NO!"*

When he got individual voice lessons with no chance of encountering singers like me, he flourished.

* * *

The day Gus discovered Laurie Berkner I remember thinking, *Oh, thank God, I won't have to listen to Barney anymore.* There she was, every day, on Nickelodeon, all boingy red curls and mismatched neon jeans and tops, a look that complemented her traditional folkie voice. The songs tapped into the things the littlest kids loved—being chased ("I'm Gonna Catch You"), their fascination with their own babyhoods ("Five Days Old"), their worries assuaged (the dark is not so frightening once you hear "Moon Moon Moon" since now you know the moon is "your nightlight"). He was hooked. So was I. There are several songs that are meant to be happy, but bring adults to tears. Henry enjoys making me cry by playing Laurie's "My Family" ("When you're in my heart / you're in my family") juxtaposed with all different kinds of family configurations—go ahead, put this book down and YouTube it now. I dare you to get through it without blubbering.

The first time Gus felt the need to be closer to Laurie, he was five. Laurie was staging a toddler Woodstock in Central Park, and we were there with Gus's then love interest, Tressa. At that point in his life, Gus had no compunction about wandering away; the caution that hits most kids around two or three, that it's safer to stay near the Big People, had not occurred to him, and wouldn't for many years. This is why Gus had a rope tied around his waist. Tressa's father, a man with the same luminous, sad eyes as his daughter, was not the kind of person to judge. As we spread our blanket out in the park, cheek to jowl with other kids and parents, I sat on the rope so Gus couldn't go anywhere. But he had enough room to bounce around and was happy. The concert over, the "Good-bye" song sung, I went to pack up our stuff, noting a little wistfully how Tressa clung to her father's leg. In that moment that I stood up, Gus was gone.

Thousands of kids, acres of meadow. Why didn't someone stop him immediately?

"What does he look like?" the policeman asked as I hyperventilated.

"Well, he's got straight brown hair and is wearing overalls and, oh yeah, *he's dragging a twelve-foot rope behind him.*" How hard could it be to find him?

At that moment I answered my own question. "I know where he is," I said, and the policeman sprang after me. "Can you get me backstage?"

When I got backstage, Gus was eating an apple with a four-hundred-pound tattooed roadie.

"Huh, I thought that rope was a little weird," the guy said. "But he was back here, so I figured he belonged to one of the musicians."

Soon after this incident, I wrote an article about Laurie so I could score good seats and make sure Gus got to meet his hero. That ploy worked a little too well, because he then wanted to meet her after every concert he went to for the next ten years. And he did.

One day, when he was around eleven, we were walking around our neighborhood when a mop-topped redhead shouted, "Hi, Gus!" and Gus shouted back, "Laurie, hi!" and we went on our way. I realized at that point that Laurie Berkner tended to cycle through fans, so if the older ones were six or seven, maybe Gus's age made him memorable. Well, that and the hopping. He would worm his way into the front row of whatever show she was doing and become the Human Pogo Stick. It's not the way most of us hop, with just our legs. Gus throws his head back and closes his eyes, as if looking at the source of joy would just be too much for him; then he bounces with such force he practically levitates. It's pretty hard to miss.

After one concert he managed to barge his way into her dressing room. Mortified, I yelled at him to get the hell out of there, and I heard Laurie serenely saying, "Hey, Gus, I just need to change and I'll be out in a few minutes."

There is no celebrity easier to love.

* * *

A bedtime conversation, when Gus and Henry were ten:

> HENRY: Girls want to hold your hand. I'm never getting married.
>
> GUS: I'm going to marry Laurie Berkner.
>
> HENRY: Gussie, I know Mom says old women are good, but she'll be like seventy when you can get married.
>
> GUS: *I am going to marry Laurie Berkner.*
>
> HENRY: YOU CAN'T.
>
> GUS: [bursts into tears]
>
> HENRY: Wait, wait, let me think . . . Oh, she has a daughter.
>
> GUS: [sniffling] Lucy?
>
> HENRY: Lucy. Maybe Lucy can sing.
>
> GUS: [brightening] I'm going to marry Lucy!

* * *

One day recently I decided to call Laurie. This is not like getting Madonna on the phone, though it's harder than it used to be. Until a few years ago, the forty-eight-year-old singer had a listed phone number, but then one too many hedge fund managers called to see if they could get her to play at their three-year-old's birthday party. Anyway, I reached out to her because I had to know: Was Gus the only one? Did she have other fans this long in the tooth, desperately clinging to the shores of their childhood by faithfully attending her shows when their peers had all become Katy Kats or Swifties? And were there perhaps other Flappers in her fan base?

"Oh, Gus is *totally* not the only one," Berkner said. "A bunch of kids on the spectrum are obsessed with me. I wish I knew why. I have been trying to get more insight into this myself, because whatever I'm doing, I'd like to do it more."

Berkner directs me to a blog called Autism Daddy, where the dad describes his nonverbal twelve-year-old sleeping with a framed photo of Laurie like it's a teddy bear. A mom of an autistic child told Laurie that her son's first word was "pig" because he wanted to put a pig on his head at one of Laurie's concerts. Prior to that, "he had never wanted anything badly enough to ask for it verbally."

There is something about Berkner's music—its simplicity; its clear, focused ideas—that makes it immediately memorable. And in her own life, she appreciates simplicity and repetition. "I'm a huge Philip Glass fan, and I love West African music that does the same thing over and over again. A lot of kids find that kind of music comforting. But maybe autistic kids even more so." Before she started writing for kids, Berkner played in an all-women rock band—and, for her day job, worked at a program for profoundly autistic adults. Some were violent; one would spend all day turning in circles. They had their musical obsessions, too. "One guy would only listen to Gil Scott-Heron, and another, only Spanish music radio programs. He listened through headphones and wore them most of the time. I never forgot how powerful that experience was."

On some level, Berkner says she understood. "My parents both worked and I didn't see them until late at night," she says. "And even though I had a brother, I remember feeling lonely a lot. I would do all these very repetitive or very weirdly focused things. Like, I would stare at one diamond-shaped-pattern fence at my school, cross my eyes, wait for the diamonds to pop out. All these habits that must have looked strange to other kids, but they helped me soothe myself."

Berkner's other little secret? "I don't listen to music that much. It's too much stimulation, and I can't take it in. Maybe that's why I identify a little bit. When I find something soothing, I'll listen to it over and over." And that, Berkner says, is how she writes for kids: she goes back to the person she was, the anxious little person, and asks, *Would I have liked to listen to this?*

"For those of us who are alone in our heads a lot," Berkner continues, "music can draw us out."

That's what Berkner's concerts do—for everyone, of course, but maybe a bit more for someone like Gus. When he was little he wanted to spend all his time alone. He couldn't have a real conversation with other kids. A Laurie Berkner concert was his first chance to be a full and equal participant in a shared community.

* * *

Here's another thing about music. It's obviously an emotional wellspring to most of us. But for autistic kids, it can teach them about *other people's* emotions.

One outgrowth of the autistic person's lacking the "theory of mind" is that many cannot read other people's facial expressions. When Gus was little and had done something to make me crazy, I spent a lot of time pointing at my head and shouting, "SEE THIS FACE? THIS IS NOT A HAPPY FACE," and he would cock his head like a Scottish terrier, trying to suss out what my face meant. Eventually he learned to distinguish Happy, Sad, and Angry. But think how many expressions there are. He could no more tell you what frustrated or wistful or jubilant looked like than he could tell you the square root of pi.

And that's where music came in. Because music does one thing for kids on the spectrum that words often can't do. Geraldine Dawson, a psychiatrist and the director of the Duke Center for Autism, studies the effect of music on the brain; she explained it to me one day. "You know how so many kids on the spectrum love Disney movies?" Dawson says. "Everyone has been trying to figure out why. But we think it's because the music in the movies gives the kids emotional cues—cues that they wouldn't pick up just by looking at people's faces or listening to what they said."

Dawson's theory seems so right. At a certain point Gus became obsessed with the song "Poor Unfortunate Souls" in *The Little Mermaid*. This is the point in the movie when the octopus witch Ursula, who is willing to give Ariel a human form if she forsakes her beautiful voice, is singing about all those creatures who have given voices to her. Gus would sing that song again and again and again, turning to me, with eyes shining bright, explaining: "Ursula is a *villain*, ha-HA." She was a safe introduction to the idea, previously completely missed by him, that other people could be "in pain, in need"—and that there were some people (or octopuses) who *enjoyed* that. Ursula was Evil, neatly packaged and explained. Once the music cued Gus in to evil, there was sort of a cascade effect in recognizing the signs that a person may be up to no good.

Around the time "Poor Unfortunate Souls" was in heavy rotation and Gus was wanting to wear his Ursula costume to school every day (the eight legs were cumbersome), I found him studying a website called Evil Eyebrows. It featured images of the Joker, Scar in *The Lion King*—and Jack Nicholson in *The Shining* (which really could have been Jack Nicholson anywhere, I suppose). "Yup, those are evil eyebrows all right," Gus said with satisfaction. "You see, Mommy?" And then he gave me his best Evil Eyebrows, which were really more like Groucho eyebrows, but never mind; he was practicing connecting a facial expression with an emotion. An emotion that is far subtler than mere Happy, Sad, or Angry, and therefore very hard for him to see. It was the facial-recognition equivalent of the SATs. He is still getting there, but it was music that sent him on his way.

Whenever I think I can't stomach one more minute of Gus's constantly repeating the same videos and music on the iPod and on YouTube, I think of my conversation with Geraldine Dawson. The fact that Gus can control the movie on YouTube or Netflix—can stop it, rewind, play it over and over again—or can stop and replay music on my iPod so many times it makes my ears bleed—all this seemingly annoying behavior means he can take in the world on his own terms and at his own pace. Starting and stopping, breaking a song or a movie down into notes or frames, repeating ad nauseam—it all looks insane. But these damn screens and machines give him entrée into all the communication we take for granted. The screens may not be real life. But just maybe they are providing scaffolding to help him create that life.

* * *

To this day, Gus's everyday world is defined more by music than words. If I ask him to do something, he may ignore me; if I sing that same request, he obeys (even when I'm out of tune, which is pretty much all the time). Sometimes he seems to have a form of synesthesia, a confusion of sights with sound. I realized this when we were walking to school recently and he saw a rainbow.

"MommyMommyMommy, look!" he said, pointing at the sky. "It's a major-chord day." Dark, rainy days are minor-chords days, and if he knows something fun is going to happen at the end of the day, the day is a *crescendo.*

John's dreams for Gus's future have always centered on music. (And for a while there, Henry's career centered on being Gus's manager.) When Gus took singing lessons, he was a perfect boy soprano, and his teacher wrote me gushing notes about his musicality. The problem was that he couldn't look at the audience, so his performances were facing backward while he hopped. This worked pretty well for Pinocchio's "I've Got No Strings (To Hold Me Down)" but not much else. When Gus could finally move his fingers separately and thus take piano lessons, I was shocked to find him listening to songs online, then going over to the piano and playing them by ear. He's not a savant—he makes mistakes and takes a few tries to get it right—but he can do it almost effortlessly, and with great feeling, too. In fact, too much feeling: He sat down to play "Scarborough Fair" after he heard it, and made himself cry. Then he did it again, and cried again. And now he won't play it at all.

And therein lies the problem. It doesn't matter how good he gets; I can't imagine him performing in any way. Or, rather, before he does, he has to have that thing he has yet to develop, that theory of mind, so that he understands he is doing this for others, not just himself. You can't be a good performer if you haven't mastered the concept of audience, of playing for the enjoyment of others.

But really, who cares? He revels in the music for himself. At night he'll sit down and play the eclectic collection of pieces he loves: "Für Elise," Disney songs, Lady Gaga, the Beatles, scary music in horror movies (he enjoys those notes of warning we get just before the severed head plops out of the closet). But I've noticed this: he won't play Laurie Berkner, ever. He can listen to her all day. He knows every song. But he will not even attempt to play her, despite my cajoling. I ask him why, and he just shrugs.

I have a theory, though. I think it's the same reason he wants to watch the buses go in and out of Port Authority, but doesn't want to ride on them. Some things in life are perfect just the way they are.

Five

VROOM

I can't stand it. He did it again.

There he is, doing the perp walk. Head bowed, hands cuffed behind him. Another poor black man arrested in his ill-fitting Nike-knockoff windbreaker.

Only it wasn't guns or drugs. And this wasn't just another lost soul to me. Every time I look at Darius McCollum, I'm reminded how innocence can be misunderstood and twisted, and worry for Gus's future.

On November 11, 2015, Darius McCollum walked into Port Authority, the United States' busiest bus terminal, and took a bus. He got behind the wheel, drove it out of the station, and motored through Brooklyn. Then he got caught. And now he's behind bars. Again.

A bus is generally the last thing most people would want to steal from Port Authority. (I'd opt for Cinnabons.) But Darius is not most people. He was fifteen when he began his career as a serial transportation thief, impersonating an engineer and driving an MTA subway for six stops before getting caught. Well, that was cheeky for a teenage boy, and Darius became a local hero. He grew up in Queens, in a modest family; he had a fascination with trains, planes, and buses from the time he was a little boy, and by eight had memorized every stop on the New York City subway system. Large and lumbering and decidedly one-note in his interests, Darius was bullied at school; he sought refuge at the terminus of the F train near his home. There the MTA workers, charmed by the smart young teenager, taught him everything they knew. He was a good student. A little too good.

Darius is now fifty, and over the years his habit of purloining public transportation became less adorable. He's been arrested twenty-seven times—and those were only the times he was caught. He has probably stolen buses hundreds of times, because when he makes off with a bus full of passengers, he doesn't exactly terrorize them. What he does is drive them on their appointed route, very politely and correctly announcing the stops. Nobody knows they've been hijacked.

That Darius is on the spectrum should be a surprise to approximately no one. For the crime of loving public transportation, Darius McCollum has spent over a third of his life in jail. Even when he's already in deep trouble, he finds a way to sabotage himself. A few years ago, he was living in a new place and told some acquaintances how he could steal a bus nearby. Unfortunately, that place was Rikers Island—there are buses there to transport inmates—and the people he told were his jailers. And that is how a man who never hurt a single person in his life became a flight risk with restrictions on his incarceration greater than your average murderer.

That was in 2013. McCollum had been free for less than two years. This time, when he was caught, he had a fake Homeland Security shield and gave officers the ID. McCollum was charged with grand larceny, possession of a forged instrument, impersonating a police officer, unauthorized use of a vehicle, possession of stolen property, and trespassing, police said. At his arraignment several days later, his attorney, Sally Butler, said what anybody who has spent thirty seconds with an autistic person was thinking. "If Darius can walk onto a bus, they should hire him to teach them how to catch terrorists . . . Why not? Let one hand wash the other. If anyone can just walk into Port Authority and steal a bus, you think maybe we need some assistance?" When I read this, I thought about the 2002 Leo DiCaprio movie *Catch Me If You Can*, based on the true story of Frank Abagnale. Before he turned twenty, Abagnale had impersonated so many people and pulled off cons worth so many millions of dollars that finally the FBI hired him to help catch other check forgers. Admittedly, there isn't as big a market for transportation thieves as con men, but surely there has to be a way for the feds to put McCollum's skills to use.

A few months later, from his cell at Rikers Island, McCollum asked for help. He explained that the obsession that had put him behind bars for half his adult life was out of control. "I can't seem to get myself out of this on my own," he said in an interview with the Associated Press. "But what am I supposed to do? There's no AA for buses or trains."

* * *

Many, many little boys go through what I think of as the Choo-Choo years: usually sometime between two and seven, trains (and often planes and buses) become everything to them. Then the passion ends.

But for many people with autism, it never goes away. In fact, it often intensifies.

Certainly it did for Gus. For the first year of his life we couldn't travel by subway because Gus would shriek from the moment we walked onto the platform. But that agonizing fear of the noise and action turned within a year to rapture. The first sound he made was not "Mama" or "Dada," but the "bing . . . bong" that warned of the subway's closing doors. Before he could form his own sentences, before he would even ask for milk or juice, he would turn to me and say, apropos of nothing, "Stand clear of the closing doors, please." Later, when he began watching YouTube, that directive would be repeated in its various permutations around the world: UK, "Mind the gap"; Germany: "*Türen schliessen*"; Japan, "*Happoufusagari!*" When he and Henry were three, their *Thomas the Tank Engine and Friends* toy trains were marked with bits of red and yellow tape, respectively. Gradually all the trains were tipped with yellow, as Henry became adept at ripping off the red tape and marking all the trains as his. But when Henry's train love gave way to Power Rangers at five, Gus got them all and continued to expand his collection. Today, we have all ninety-nine characters (or is it a hundred?) and many multiples. He will not give them up. When he needs to unwind, he stims with them, making that click-clack sound trains make. He's confined to doing this in his room because the sound is as annoying to everyone around him as it is relaxing to him, and frequently I ask if he's ready to donate the trains to a younger kid. "I will," he says. And then, after three seconds of thought, "But not now."

Researchers have looked at why *Thomas the Tank Engine* (a series of British children's books about talking trains that morphed into a television cartoon series, a line of toys, and the poster boys—uh, *trains*—for the UK National Autistic Society) is so deeply appealing to autistic kids. It starts with the train's faces—not just the easy-to-read expressions, but the very fact that the trains *have* faces. It's the melding of machine and feeling that seems so core to the autistic self. The personalities and characteristics of each character are unchanging; Gordon is always going to be the fastest and most powerful engine, Edward will always be a friend to all, and Thomas will always be an overzealous little twit. (Not that I've overthought this.)

There is also the static background and scenery. While the narration must have cost considerable bucks—the cartoon was narrated by Ringo Starr in the UK and George Carlin in the United States—the Thomas animation is very cheap. This means there isn't a lot going on besides the trains in the foreground. Even the faces of the trains don't have much detail—they are happy or they are sad, without a lot in between. This is extremely soothing to people who can be easily distracted by tiny details other people wouldn't notice, or who find the subtlety of human expressions hard to parse. In a world where emotion could be so confusing, nothing could be clearer than the slightly creepy mad or happy faces on a Thomas or a Percy—and they get mad and happy with great regularity. Their emotions are almost binary. Because in the land of Sodor, things go wrong, and then by the end of the episode they're set right. Always.

And then, there's the trains' suitability for detailing and classification. The original tiny wooden trains each cost fifteen to thirty dollars, which is, I believe, the fault of autistic people. Why? Because they notice knockoffs immediately. Don't mess with autistic kids and their originals. The auction houses of Sotheby's and Christie's may have extremely well-groomed, socially adept people as their front men, but I'm convinced the people who really know what they're doing are people in Lycra golf pants with their shirts half unbuttoned, railing about how anyone can see that the splatter patterns in this so-called Jackson Pollock are not authentic.

For many years, *Thomas the Tank Engine* dominated our lives. There were years of *Thomas*-themed birthdays, movies, books, videos; entire vacations planned around seeing a life-size steamer Thomas train. Somewhere there exists a piece of *Thomas*-related pornography I wrote to my husband. Wisely I didn't show it to him, but it involved Gordon (the largest and strongest of the engines) and Emily and whistles and smokestacks and engines being coupled and a satisfied Emily, smoke feathering out of her stack, murmuring to Gordon, "You're a really useful engine."

Eventually Thomas gave way to replicas of New York City subway cars; Gus has every model of those, too, and eagerly awaits the arrival of new ones. The new Second Avenue subway line in New York City, which will eventually stretch from Hanover Square to 125th Street, is every bit as significant to him as Comic-Con is to a Trekkie. For years Gus attended a clever program in New York City called Subway Sleuths, an after-school program for train obsessives on the spectrum who (so the thinking goes) may be more inclined to learn the rules of social interaction when they are practicing them around a shared interest. "We weren't trying to get kids to be neurotypical, but to get kids to communicate in whatever way they do," Susan Brennan, one of the creators of the program, told me. "So our focus isn't on social skills; it's on building connections and awareness that there are all these social rules. That's an important step—just being conscious that there *are* all these rules. Some kids will be able to put them into practice better and more quickly than others, but everyone is more likely to be receptive to social cues when they're doing something they love."

At first I thought Subway Sleuths was silly, just another way to occupy the kids for a little while so that parents could have a break. Then one day I asked Gus about another kid, Lev, he met in Sleuths. "What do you talk about with Lev?" I asked. "Oh, timetables, or the 1, 2, and 3 line, the weekend changes on the B and D. You know, Mommy," he added, "things that are important."

It is a trope of autism that people with the condition are

not as feeling as neurotypical people. That couldn't be more wrong. It's just that sometimes they have deep feelings for things the rest of us don't. The parody site the Onion News Network, for example, has a series of news segments from a man they call "the Autistic Reporter." Here, Michael Falk (played brilliantly by the actor-playwright John Cariani) sees the news somewhat differently than the average person. In a news report called "Train Thankfully Unharmed in Crash That Killed One Man" Falk notes that a "hundred-thousand-pound CometLiner 2 stainless steel car" ran over a man who had jumped on the tracks to retrieve a woman's purse. He was instantly killed—but "luckily," Falk adds, "there was no structural damage to the car's chassis, so it was only a matter of cleaning the train to remove the human debris to return it to its pristine state."

Now, I'm not saying Gus would be immune to the death of a human. But he'd be truly delighted that the train was not harmed.

These days, after he does his homework, Gus and Michelle (his current caretaker/train buddy) head to one of his favorite places: Port Authority, Penn Station, or Grand Central. At Grand Central he is greeted by many of the conductors; one had an MTA badge printed up for him, and another gave him his conductor's cap. Many let him get in their booths, flip on the microphone, and announce the routes, which of course he knows: "Harlem–125th Street, Melrose, Tremont, Fordham . . ." By the time he makes it up to White Plains and North White Plains the passengers often end up clapping.

A few, of course, get irritated. He has had anxious passengers check the booth to make sure he's not actually *driving* the train. Last year there was an incident where a conductor on the New Haven line messed up the order of the stops and he corrected her. When she first ignored him and then gave him the stink eye, he sobbed. She smirked. To that cow, I wanted to say: "Sure, he was crying because you wouldn't speak to him. But mostly he was crying because by not announcing the stops and connections correctly, you were dishonoring the train."

* * *

It is my fear that for all his passion for vehicles, Gus will never be able to go anywhere by himself. Or, alternatively, that he will go everywhere himself, and disaster will ensue. Just as it has for Darius McCollum.

Thankfully, at fourteen, Gus is delighted to just watch the trains and buses and announce their routes. He shows no interest in actually being a driver. He seems less likely to become a train thief than a trainspotter. Trainspotting is a phenomenon that began in the UK in 1942, when Ian Allan, a kid in the press office of the Southern Railway in the UK, got tired of answering endless questions from train enthusiasts about the locomotives. He suggested to his office that they put out a simple booklet detailing the trains' vital statistics. His boss thought he was insane, and so Allan produced the book himself. *The ABCs of Southern Locomotives* sold out its first run immediately. Further guides to every railroad line in the UK followed, and trainspotting clubs (then called loco-spotting clubs) sprang up everywhere. By the late '40s these clubs had a quarter of a million members. In the '50s and '60s a million ABC guides, listing twenty thousand locomotives, were being sold every year. Ian Allan became wealthy.

But with the death of the locomotive as a main form of British transportation, membership in these clubs waned. Now they have a mere ten thousand or so hard-core members. Still, England has always embraced its eccentrics. Go to any station in the UK, and you'll see a few of these anorak-clad outliers jotting down the numbers of passing locomotives in their well-worn notebooks. They look solemn yet content. Years ago Chris Donald, the founder of the popular British comic magazine *Viz* and an enthusiastic trainspotter, told an interviewer, "In some ways you can get as much from a train as you can from a woman." No word on how his wife, Dolores, mother of their three children, feels about this.

* * *

Given that Gus doesn't seem headed for a life of train hijacking, why do I get so worked up about Darius McCollum?

Because I have seen what happens when Gus has a compulsion. I've seen the hours and hours he devotes to watching the trains, memorizing them, knowing the routes, learning the names of the drivers, learning where *they* live. What if that love turns into something else? If he decides he wants to drive a train instead of watching it, he will be driving a train.

Darius McCollum keeps me up at night, so I started a page on Facebook: "Darius McCollum Needs a Job." I needed to know: Why for the love of God couldn't the MTA find work for this man?

When I was obsessively discussing Darius on my Facebook page, one woman wrote, "Well, of course the MTA couldn't hire him; their insurance will never stand for it. He's just too unpredictable." And I wanted to shout at her, you couldn't be more wrong! Darius McCollum is 100 percent predictable. Let me tell you what he's going to do as soon as he's free: *he's going to steal a vehicle and safely drive it around.*

One woman, happy that I had started the page, wrote to me privately. Ramona A was thirty-three years old in 1983 when she was hospitalized for a serious eating disorder. McCollum, who had just stolen his first train, had been committed to her unit. "He was this extremely gentle, sweet boy who had great difficulty communicating, couldn't relate at all to his peers (the other patients viewed him as 'weird' and he was shunned), and could only talk about the subway system. He had all these odd and off-putting behaviors which, looking back, I can now identify as putting him on the high end of the autism spectrum. Like he would just talk and talk and talk at you—usually about subways and trains. His speech was very fast, and often garbled. If you managed to get a word in edgewise, he would simply ignore you and keep talking. He would invade people's personal space. He was completely oblivious to social cues. Also (sorry—this is gross) he would sneak around the unit until he found a place where there were no other people—and he would poop on the floor. I'm not sure why. Maybe that was his way of showing how angry he was at being there." His mother would visit him, but she seemed at a loss. But he was such a nice person, Ramona recalled. "I just can't imagine him being in prison."

Ramona's note upset me so much that I made the mistake of visiting Darius's own Facebook page. There is a photo of him happily posing in front of a D train; and there are several recent posts:

I need a wife. Looking for someone to care for, love and be able to share myself with and also someone who can understand me for who I am. I just want to be loved.

By curiosity, I was wondering if there are any people out there who may like trains or are even just train buffs. It's a passion of mine.

I have and always will be a one women man. Even though I am looking for someone, I just want that one person.

Then there was the most recent post, a photo of a woman with long black hair. "Now you know, this is my current young lady who I am choosing to be with and I love her very much." That was November 7, 2015. He was arrested for taking a bus on November 13.

I wrote to Darius's "current young lady," Mary. She is from the Philippines, and she and Darius met on some sort of dating site. They have never met in person. She seems very far from the "mail-order brides" from the Philippines you've heard about. She had become very attached to Darius, and was deeply upset by his arrest (note: English is obviously not her first language): ". . . when I saw him he arrested I am shocked because he have many secret of his life that he hide to me all, now I am much sad and worry to him, hoping someday that he do good for better person. I did not judge him, I understand him." But when she discovered that Darius was not, as he had told her, an MTA employee, she was less understanding. She felt betrayed, and at a loss. Mary had never heard of autism before.

I emailed Darius's attorney, Sally Butler, offering my help without exactly knowing what that help would be. I think it involved winning the lottery; that way, I could buy Darius a "shadow" for a year, someone whose job it would be to make sure he was safe doing *his* job, thus convincing the MTA to take him on.

I was happy to hear that there has been an enormous outpouring of sympathy for Darius. There was a documentary being planned about his life, and a fictionalized movie, starring Julia Roberts as his lawyer.

Still, what good did it do? He was in solitary confinement at Rikers Island. He is an only child, his parents are elderly and live in South Carolina, and at any rate they had given up.

"I think my office and our team are pretty much his family at this point," Butler said.

I blubbered for a while, and then I called the MTA. I needed to know. What would be SO terrible about finding this guy a job?

Adam Lisberg, the external director of communications, is trying to be patient, but he talks to me like I'm a slightly dim toddler who needs a time-out. "Ummm . . . he stole a *train*. He steals buses. He will never be hired here in any capacity. Over and over and over he has failed to follow MTA laws." The MTA does not take a position on whether he should be helped or criminally prosecuted; they just know there is no place for him in their system. "Do you think we should have someone who has impersonated an MTA authority actually work for the MTA?" Lisberg asks.

Yes. Yes, I do.

Lisberg explains that the problem is not that Darius is autistic; in fact, he suggests that were it not for people on the spectrum, there might not *be* an MTA. "There are a fair number who are on operations side, or are planners with the buses," he says. "They love it. They can't get enough. We have people all over the system who are on the spectrum. That's not it. But this guy can't control his impulses. If he can't control his impulses from the outside, what do you think he'll be like if he actually works here and gets it into his head that he wants to take a particular bus or train that day?" Lisberg can see he isn't convincing me. "Look, bus drivers can lose their jobs if they can't keep their uniforms in good shape. I think stealing a bus goes beyond that, don't you?"

No, Mr. Lisberg, I do not. What I think is that Darius McCollum—and someday, my Gus—will be the best employee you ever had.

And their uniforms will be spotless.

Six

BLUSH

"*Don't . . . do . . . anything*," Henry hisses as I continue to do exactly what I was doing before: nothing.

In the state Henry's in, it's best not to establish eye contact. So I continue to answer email on my phone as I mutter, "What exactly do you think is going to happen here?"

"I know you," he says. "You're going to speak to him. You're going to ask him for a photo. You're going to *dance*."

Thanks to my friend Janice, who ran *Billboard* and *Hollywood Reporter*, Henry and I are at a photo shoot in LA for one of his cultural heroes, Andy Samberg. Samberg and his crew, Lonely Island, have a movie coming out about an obnoxious Justin Bieber–ish rock star. The premise of this shoot is that he and his boys are emerging from a Hummer (cue fog machine) with their entourage of bodyguards and a driver. Only the bodyguards are seven-year-olds and the driver is Henry. A few days earlier I'd gotten a call from the photo director of *Billboard*.

"Can your son act?" she asked me.

"Not at all," I said. "But he's really good at staring straight ahead and not smiling."

"Fine, he's hired," she replied.

So I used frequent-flier miles to get to Los Angeles so Henry could meet Samberg. And I am having a proud-mother moment: he looks splendid in his tailored black suit, Dolce & Gabbana loafers, sunglasses, and fake CIA-inspired earpiece. He even enjoyed being fussed over by the stylist, or enjoyed it as much as his general gloominess would allow.

The problem, for Henry, was that at any moment I might do something that every single person in that room would be sniggering about later. Like tell him he looks nice. Or ask Samberg for his autograph. Or take a photo of the adorable prop girl he told me was so pretty. OK, maybe I did that. Maybe he died internally. But: memories!

It is thrilling that someone in this world thinks I'm a loose cannon. I am about as loose a cannon as Emily Dickinson. But in Henry's mind, I am only about thirty seconds away from grabbing Andy Samberg and making out with him on the hood of the Hummer. Just because one time, one time only, I sang and spelled out the words to "Y.M.C.A." while his bus drove off to camp. I mean seriously, if you came of age in the 1970s, wouldn't you?

Here's a dirty little secret of parenthood: making our children cringe is one of our great pleasures. Yes, sometimes the embarrassment is inadvertent. As a physician, my mother believed all medical details were inherently interesting, which may be why she used to love regaling my friends with stories of how difficult my birth was; then she would offer to show them her cesarean scar. But more often, as parents, we know what we're doing. Embarrassment is muscle flexing with humor. In an interview with *Entertainment Tonight,* Michelle Obama noted, "Barack and I take great joy in embarrassing our children. We threaten them. Sometimes when you see me whispering to them in a crowd I'm saying, 'Sit up or I'm going to embarrass you. I'm going to start dancing.'"

Embarrassment works particularly well from the ages of, say, twelve to eighteen. Henry has a history of not texting me when he arrives at or leaves a place. So recently, when he was going to an evening baseball game with a friend, I informed him that if he didn't call at a certain time, I had gotten the number of the Mets announcer and he would hear over the loudspeaker at Citi Field, "Henry's mother wants him to phone home."

This might not work in a few years. But he is only fourteen, and there are still fumes of my all-powerfulness. Maybe I actually managed to get the announcer's number. Maybe I'd make that call. Could happen. Scary thought. He still believes. It was like the time he was six and he asked me what part of the buffalo the wings came from. *Huh. Well, if Mom says they can fly, they can definitely fly.*

In my defense, Henry has devoted a good portion of his life to embarrassing me, too. I kept a note from his beloved fifth grade teacher, Ms. Wahl, who I thought of as the Most Patient Woman in the World: "Hi—Henry is refusing to do the Pledge of Allegiance at graduation practice. He says that he doesn't agree with America and that his father is a socialist. I said if he has a note from home I can excuse him but I kind of think he should just participate and say it. Thoughts?"

Being embarrassed seems terribly unpleasant, but like with many unpleasant things, we never stop to think of how important it can be to our humanity. If you are embarrassed, you understand certain hidden social rules. You know they've been transgressed. An overreaction to other people's behavior at fourteen means that you are learning, gradually, how to modulate your own.

But what if you have a child who cannot be embarrassed by you—and doesn't understand when he embarrasses *you*? What then? Nothing makes you appreciate the ability to be embarrassed more than having a child immune from embarrassment.

* * *

Recently I read this headline: "Philly Mom Gets Nasty Anonymous Letter about Her Son with Autism." I winced, imagining how bad this was going to be. It was worse than that. Bonnie Moran, a woman with an autistic child, woke up to find this letter in her mailbox (not copy edited):

> *To the parent of the small child at this house,*
>
> *The weather is getting nicer and like normal people I open my windows for fresh air. NOT to hear some BRAT screaming his head off as he flaps his hands like a bird. I don't care if it's the way you raised him or if he is retarded. But the screaming and carrying on needs to stop. No one wants to hear him act like a wild animal it's utterly nerve wracking, not to mention it's scaring my Normal chil-*

dren. By you just standing there talking to him don't do anything. Besides you look like a moron as he walks all over you. Give him some old fashioned discipline a few times and he will behave. If that child needs fresh air . . . take him to the park not in out back or out front where other people are coming home from work, have a day off, or just relaxing. No one needs to hear that high pitched voice for hours. Do something about that Child!

Moran cried for hours.

This story has a happy ending, though. I contacted her after reading the story; she told me that eventually she found out who sent the letter, and invited her to come and spend time with her son to get a better idea what autism is like. The neighbor doubled down, saying that Moran was a horrible mother who was only trying to get attention. But when Moran posted the letter on a local Facebook group, she got many playdates for her son from appalled neighbors who wanted the boy to feel welcome in the community.

This story reminded me: all mothers of spectrum kids have moments of mortification. Me, too.

On the one hand, I'm blessed: when things don't go his way, Gus doesn't have meltdowns. On the other hand, even without them, social norms are meaningless to him. "He likes the MTA a little bit more than the rest of us," I'll say as Gus forces some unsuspecting out-of-towner holding a map to listen to his subway directions. I have dragged him away from conversations about God with various homeless people after he's demanded I give them my money, and I have been asked to quiet down in movie theaters and plays because he didn't understand what a whisper was. Halloween is the best holiday for us, and indeed for most parents of autistic kids, I think. There is nothing your child could do that would be too weird. Though he doesn't eat any candy, Gus loves collecting it. It's the perfect amount of human engagement for him: you say one phrase to people at the door, people admire you, you move on. (At least now he does. He used to storm their apartments and refuse to leave until he'd buzzed through every room.)

Last year, at thirteen, Gus was Maleficent, complete with flowing robes and horns. He knows she is a girl and does not care. She can transform into a dragon, so it's all good. Henry, who dressed as a Corinthian or something (I never quite got this right, but we shopped all over for historical accuracy), was mortified that people were taking photos of his brother as he threw his head back and MWA-HA-HA-HA'ed at the top of his lungs. "Sweetheart," I said as Henry attempted to render himself invisible, "this is why we live in New York."

Modesty is also an entirely foreign concept to Gus. As someone who wouldn't go to the bathroom in front of a dog, never mind another human being, I am rattled by a child who doesn't understand the point of closing the door. Gus never notices that his pants are riding so low that his butt is showing, nor has he learned, even at fourteen, that when there's company, it's not perfectly fine to walk to the shower naked. Or, rather, he understands that he's supposed to wear a towel, but only because I say so. He's still unclear about where the towel's supposed to go. Generally he slings it over his shoulders.

"Aren't you embarrassed?" Henry says to me when we're walking down the street and Gus is quietly quacking under his breath. Henry reminded me that for the past year Gus had been wanting to walk to school by himself, which to him seemed perfectly reasonable and to me seemed like inviting him to play his own personal game of Frogger.

"I mean, imagine if he was doing this while walking on his own. GUS, CUT IT OUT," Henry shouts, for the hundredth time that day. When he's feeling a bit more hopeful, Henry has a theory. "Imagine if thirty years from now we find out Gus was faking everything and that he was actually a British mastermind trying to infiltrate our family."

* * *

There are endless studies on embarrassment. (And some are kind of fun: What happens when you ask a subject to stare at a variety of people in photos and then tell them that according to eye measurements, they've spent much more time than the average person staring at the people's crotches? Hilarity ensues.) But in general, embarrassment is a social emotion: we feel embarrassed when something we do, or something someone else does, conflicts with our image of ourselves in front of a group of people. The key phrase here is "image of ourselves." If one of the primary manifestations of autism is the inability to understand that other people have thoughts, feelings, emotions, and needs different from ours, then it makes sense that many aren't self-conscious; they don't have a sense of who they are in relation to other people. Certainly Gus doesn't.

So what does a parent do? On the one hand, you try to control the most socially unacceptable behavior. "I only can touch myself in the privacy of my own room!" Gus has announced to me on several occasions, and while I take this as a hopeful sign that my message has sunk in, I also have to pray that he does not see this as an interesting conversation starter at a friend's birthday party.

Then there are lots of other merely annoying or clueless behaviors that I haven't been able to eliminate altogether, but have occasionally been able to put to good use. For example, for years I couldn't get Gus to stop answering the phone; his need to connect with people far surpasses his ability to understand what real connection is. So he would race to answer, and I would then find him deep in discussion with various people I worked for, asking them where they lived, where they were going that night, and giving them directions on how to get there. But as time wore on most work people were emailing and texting, and it dawned on me that the only people who used my home phone were telemarketers. Henry, whose entire life is devoted to pranks of one sort or another, convinced me: let Gus answer. Now, Gus patiently waits for that dead space or the recording to be finished until he reaches a live human. And that's where the fun starts. "My mom is right here. What did you want to ask her? Where do you live? What train station is that near?" At first I felt guilty, but as Henry pointed out, "Telemarketers give you the gift of wasted time, so you're just returning the favor."

Lately I've been getting fewer and fewer telemarketer calls. I suspect there's a "Do Not Call: Batshit-Crazy Kid at Home" list.

* * *

There are things worse than an autistic child who feels no embarrassment, as I found out recently. Much worse.

Gus and I were at a concert sponsored by Music for Autism, a fantastic organization that brings Broadway performers together to give hour-long concerts for kids on the spectrum. The hell with the kids; it is bliss for us parents. For an hour, we are not worrying that our kids' behavior, expressed in socially questionable ways, will impede anyone else's good time. Dancing in the aisles and singing at the top of your lungs is encouraged. In other words, at a Music for Autism concert, I am as free from embarrassment as Gus is.

At this respite from reality, performers were singing barnstormers from the Gloria Estefan musical *On Your Feet!* Gus was doing what he always wants to do, but usually can't, except at these concerts: inching closer and closer to the singer in rapt wonder, until he was a foot away, dancing with her. The rhythm is going to get you—and you and you and you, and definitely Gus.

But then, there was this child. He was about Gus's age, olive-skinned, handsome, and barely holding it together. He just kept repeating to his parents, over and over, "I'm sorry . . . I'm sorry . . . I'm sorry."

The little guy had absolutely nothing to be sorry for—except, maybe, that he couldn't stop repeating the phrase, and his parents couldn't make him stop. Did he feel sorry for something he'd done, or was this pure echolalia? (Echolalia is the precise repetition of words and phrases, very common among those with ASD, that the person doesn't always understand or mean.) I didn't know. But I know that if, for whatever reason, he lived in a state where he was embarrassed by his behavior—unable to control it while being keenly aware that it was not normal—then, oh my God, *I* was the one who was sorry. I wanted to hug him and his parents. I wanted to give him a transfusion of Gus's obliviousness. I wanted the singer to break into a rendition of Lady Gaga's "Born This Way."

* * *

Through pain there is growth. I think about this all the time. Do I want my son to feel self-conscious and embarrassed? I do. Yes. Gus does not yet have self-awareness, and embarrassment is part of self-awareness. It is an acknowledgment that you live in a world where people may think differently than you do. Shame humbles and shame teaches. One side of the no-shame equation is ruthlessness, and often success. But if you live on the side Gus does, the rainbows and unicorns and "what's wrong with walking through a crowd naked" side of shamelessness, you never truly understand how others think or feel. I want him to understand the norm, even if ultimately he rejects it.

There are signs that things are changing—though incrementally. The other day I was wearing low-rise jeans, the bane of pudgy middle-aged women everywhere. I was bending down to clean something off the floor, and I guess I didn't notice I was doing my best plumber's impersonation. With a look of infinite compassion—and using the same gesture I have used on him a thousand times before—Gus came up behind me and tried to hike up my pants.

"That looks silly, Mommy," he said as I rejoiced.

Seven

GO

"Guess what? We're going to Alaska!"

GUS: "Is there rice pudding?"
HENRY: "No."
JOHN: "How much?"

Why should I have thought this trip would be any different? That I would be greeted with cries of "YAY!" and "YOU'RE THE BEST MOM EVER!" and "IT'LL BE LIKE A SECOND HONEYMOON!"? OK, maybe that last one was a stretch. Actually, they all were.

I do not have a good history of travel with my family. At this point I don't even think of these trips as adventure so much as anthropology, a chance to chronicle the character defects of the people closest to me. And yet my romance with family travel persists. *This time it'll be different. This time will be The One.*

Always lingering in the background is my fondest wish: that Gus will become a fan of The New, or at least not its mortal enemy. It is all part of my alternative reality, where

Gus lives in Normal Land. In Normal Land, I do not have to travel with a box of Cheerios in case somewhere, somehow, they run out. In Normal Land, my Gus is interested in sightseeing and talking to the natives, rather than, say, watching the buses arrive and depart from the hotel. In Normal Land, Gus enjoys eating more than five foods. He will look at a stranger as they shake hands. He will feel when his pants are falling down and pull them up himself. And mostly, he will not cry every afternoon because he is homesick—not for his family, who are in front of his face, but for his stuff. The bedroom curtains, with snakes and lions and giraffes and several animals I think the fabric designer just made up. The little monster trucks with the friction wheels, a sound that soothes him. His villain figures, his trains, his snow globes, and always his magic staff, a gift from Maleficent, which he will still, on occasion, use when he watches her in videos, reenacting her scenes and shouting her lines the same way I used to do when I went to midnight showings of *The Rocky Horror Picture Show*. These are the things he cries for. I have taken pictures of these things and put them on his computer, and when he's feeling shakiest I show the pictures. *Look! Everything's still there, waiting for you!* On vacation his daily tears are as predictable as a thunderstorm in the tropics. They go away quickly, and he's all smiles again. But in Normal Land, Gus doesn't wake up every morning and brightly announce how many days there are before we go back home.

* * *

We've never been big travelers. For the first six years of Henry's and Gus's lives, travel consisted of the occasional overnight to a beachy place I convinced myself they'd enjoy despite the fact that they'd cling to me like baby baboons if I took them in the water. Why was their mother committing the act of swimming? While the kids around me shrieked with joy, rooting in the sand and plunging into the water, Henry and Gus would be climbing up my legs in their effort to escape the sand under their feet. Their saucer eyes and quivering lips said, "What is this bottomless abyss of dirt and wet? It is hot; there are bugs; we are being sent here because *we've done something very, very wrong.*"

Plane travel was out of the question, partially because Gus couldn't sit still, but mostly because I so resented other mothers who took their young kids on planes that I refused to join the She-Who-Is-Loathed-By-Everyone Club. I am certain that when history explains Brad and Angelina's divorce, it will have nothing to do with another woman or booze, and everything to do with frequent airplane travel with six children.

Among my most vivid travel memories on a business trip was sitting next to a woman and her eighteen-month-old. Gus and Henry were about the same age, and so, missing them a bit, I started a game of peekaboo with the little guy. The mother, delighted to have a few minutes' respite, proceeded to pound back the vodka tonics while her son menaced me with a root beer lollipop. He was determined to share, only he wanted to share it with my arm, again and again and again. I think we've established my policy about stickiness; I feel about sticky the way Donald Trump feels about the *New York Times*. You can imagine the scene with both mother and child when I finally confiscated the Hell Pop.

Not wanting to be *that* mom played a role, but so did my neurotic family. When Henry and Gus were six, I told John we were going to Disney World for the first time. John politely demurred. I think his exact words were "They steal your money and shove their bloody false American values down your throats. Is that what you want for your children?" I ended up taking only Henry. He loved the airport and plane; a beloved neighbor had just died, and he was convinced he saw Jerry sitting in the clouds. He loved the Polynesian hotel, too. The problem is, he loved it a little too much. He did not want to leave. It was not until I got to the park itself that I discovered there were two things Henry feared more than anything in the world: amusement park rides and people in character costumes. Thankfully, there was an app where you could track the movements of the characters around the park. People used this so that they could find Goofy and Donald Duck, but it was equally useful for avoiding them. The only other thing I remember from that ill-fated trip was my six-year-old son standing in front of Big Thunder Mountain Railroad shouting at random strangers, "DO NOT GO ON THIS RIDE. YOUR MOM WILL TELL YOU IT'S JUST A TRAIN, BUT SHE IS LYING. IT IS A ROLLER COASTER AND IT IS SCARY AND IT IS BAD." Essentially I paid $2,000 to glide through It's a Small World over and over. You know the little Maori child clutching what is supposed to be a boomerang in front of him? After scrutinizing it for three days I'm pretty sure it's a huge penis, and the Disney animatronic designers were having a little private laugh for themselves.

I waited another five years before taking a trip with the four of us. I decided we should go to Arizona to drink in the raw beauty of Sedona while staying in a hotel so posh nobody could complain.

On a one-mile hike to the top of a vortex rock formation, John continually expounded on the dangers we were facing: sunstroke, rattlesnakes, scorpions, dehydration. I pointed out that a one-mile trail up a gentle well-traveled slope does not necessitate survivalist skills. "OK," Henry says, "but in case we *are* abandoned and starving and we have to eat each other, I call Mom. She's the largest."

The trip culminated with me sobbing at the rim of the Grand Canyon in our ugly rental car.

"Mom? Seriously, what did you think was going to happen?" said Henry. "You brought us to the World Capital of Rocks and Pollen. Dad can't walk, and I can't breathe." John's knee problems had worsened recently, and who knew that the floating clouds of cottonwood seeds—what Henry called the Killer Fluff—would trigger the worst allergies of his life? "And look at Gus," Henry added. At the mention of his name Gus looked up from the backseat, eyes filling with tears. He hated everything about being away.

* * *

Most parents say they want their kids to have a better life than they had. But my parents had given me a great life; I just wanted my kids to have a different one. Specifically, I wanted them not to be a coward like I was. I loved sitting home with my mom and dad in their suburban cocoon watching *Mary Tyler Moore* and stealing the apple cobbler out of their Swanson TV dinners when they weren't looking. I was deliciously sheltered and cosseted. But I wanted to raise my sons to be people of the world, the kind of guys who would think nothing of striking out on their own, who didn't mind putting up with a bit of discomfort for the sake of experiencing the new and the different.

Instead, they were just like me, addicted to comfort and luxury. I have always subscribed to Joan Rivers's quip about room service: "It's like a blow job. Even when it's bad, it's good."

So our pretravel conversations would go like this:

"We need a place with an indoor pool, because otherwise all the mosquitoes will get in the pool and eat me," Henry said about one proposed excursion to New Mexico.

"Stop worrying, it's near the desert, there aren't a lot of flying bugs."

"We MUST find a hotel with an indoor pool."

"Look, this is stupid, there aren't a lot of insects in the desert. There are just scorpions."

Henry became pale. "*What?* That's it, I'm not going. This is like telling me, 'Hey, don't worry about being in that fight . . . nobody has knives, just flamethrowers.'"

* * *

When I was twelve, my mother decided to take me on a road trip in the cherry-red Buick Riviera she called the Pimpmobile. She had just discovered CB radio, and was nerdishly enthralled; as a radiologist, she decided her handle would be Hen Medic. Hen and I were setting out to see America's national parks. She let me pick the hotels, ensuring we spent much more than we could afford. My only recollection of that trip was my irritated mother screaming "LOOK OUT THE WINDOW" as I lay in the backseat, reading *Are You There God? It's Me, Margaret*. Also, there were woodchucks. These two memories cost my mother thirteen weeks of her life.

* * *

One day during winter break Henry and I were sitting in his room watching a game. "Doesn't that sound bring back good memories?" he said.

"What sound? All I hear is the sound of the heater humming."

"Yes! That. It's just so warm and it makes me think of Christmas."

"So . . . not the tree or carolers or the smell of warm cider and cinnamon? Or not one of the little trips we've taken? Not going to see the family, or parties, or—"

"I have a lot of great memories in this room," he said, a little defensively.

"If you are nostalgic about the hum of your HVAC unit, I am the worst mother who ever lived," I said.

So I tried a different tack. This time, I planned an entire trip around the concept that my husband couldn't walk very well, but also refused to acknowledge this problem. This is what cruises are for: fashioning yourself as a bold adventurer while guaranteeing you are never without a cocktail and air-conditioning.

Right before we were due to leave for Alaska, John opted instead for *his* perfect vacation: being home by himself. At least, that's how I presented it to friends. In truth, he was having minor medical issues, so I could understand how he might not want to go. But the idea of Gus on a boat with only me triggered John's worst travel anxieties.

"Watch him," John said, darkly.

It's not like John's fears were unfounded. This is because, well, that day Gus wandered off at the Laurie Berkner concert was hardly unusual; from the time he was three to about ten, on any given outing I'd usually end up describing my missing son to a concerned policeman.

It's a common problem with autistic kids. According to one study in the journal *Pediatrics*, about half of all kids on the spectrum will wander off from a safe, supervised space. There had also been the recent tragedy of Avonte Oquendo, a fourteen-year-old nonverbal autistic boy from Queens who had wandered out of his (supposedly locked-down) school, despite warnings from his parents that he was a flight risk. Parts of his badly decomposed body were found three months later, washed up on a Queens beach.

"Watch him," I heard every hour, for the next three days.

John just had to say "Paul Giamatti" to bring it all back to me. Gus was four. We were at a kid's birthday party in Greenwich Village at a party space that was a converted stable; the actor and his son were there. The party space opened onto the street. Someone left the door open, and Gus made a beeline for the exit. Giamatti tackled him before he ran straight into a car that had been backing up. I'm not sure who was wearing the Superman costume at the party, the actor or his kid, but in my mind it is always Paul Giamatti.

Gus's tendency to wander off ended when he was about ten. I was very lucky, because many parents of autistic children discover that if they have a wanderer at five, they still have one at twenty. But John remained forever traumatized by memories of Gus going missing. He also had zero confidence in my memory and attention span. Nor was he confident that Gus had finally developed a proper fear of heights, making it all too possible that he would decide that it would be fun to plunge off a deck. In this, as with many facets of Gus, John couldn't see the change. He remained certain that if he weren't there, either Gus would bolt and dive off the side of the ship, or we would go on an excursion off the ship and I'd forget to bring him back, as if he were a pair of socks.

Here are the things I learned on our trip to Alaska:

It is possible, if not at all desirable, to live in a closet with my children. That was the size of the room. When

I saw that there were bunk beds, and they folded out of the ceiling, I offered a quiet thanks to God that we weren't also sharing the room with John.

For Henry, any griping can be countered by the phrase "all-you-can-eat buffet."

Bald eagles can be as creepy as pigeons when there are dozens of them circling above you.

Dolphins really are the happiest animals on earth, or at least they seem that way when they're wake-surfing behind your boat.

The Internet is not a practical diversion on a cruise ship, which I only learned when I was presented with an astronomical bill because I had forgotten to shut off Gus's iPad.

A fourteen-year-old cannot play poker in the casino, no matter how many creative ways he comes up with to do it. It killed Henry that there was a casino on board but it was off-limits to him—because of his age, not his reputation as a cardsharp at the poker game he runs at high school lunch hour. "No, I will not wear a wire and take your instructions," I said, nevertheless slightly flattered that Henry saw his mother as a villain in a James Bond film.

My lifelong hopes that Gus was gay—what gay man doesn't adore his mother?—were briefly revived when he insisted on going to the Elton John documentary and a Stephen Sondheim revue. Cruises have daily schedules, which became a fixation for Gus: we were going to the Love Boat Disco Deck party because it was on the schedule.

Yes, he spent a fair amount of time watching people get in and out of elevators, despite the fact that that wasn't on the daily schedule. And yes, he announced every morning how many days there were till we returned. But there was no afternoon sobbing. Instead, about that time every afternoon, he would retreat to his computer and quietly look at pictures of his room at home.

* * *

The following summer I decided that I needed to adopt a different concept for family trips. First, they could not include all of us. It was impossible, as nobody liked to do the same thing. Second, they could not be about fun, for me anyway. If that's what I was promising, I was destined to lose. Instead, I had to be open to travel as fulfilling a specific goal, generally a goal it would never occur to me to have. One of my sons would have the adventure he wanted; I'd go along for the ride. Excitedly, I relayed my new plan to Henry.

"Then we're going to Caracas in Venezuela," he said. "Sure, they have the highest murder rate, but did you know they also rank number one for hottest women in the world?"

I suggested he find another goal. Which is how we ended

up in Paris during the Euro games. This time his goal was to chant for his team, England, with as many drunk people as possible. I was willing to overlook the possibility of being pummeled by angry Slovakians to spend an hour of our lives together at the Musée d'Orsay with *Whistler's Mother* and van Gogh's self-portrait. I showed him Courbet's *L'Origine du Monde* and said it lives at the Musée d'Orsay, and that being the kind of art most teenage boys can get behind, off we went to Paris.

Since I can still get lost in the neighborhood I've lived for thirty years, we picked up sightseeing guides along the way. The guides were wonderful and allowed Henry to collect interesting and entirely unrelated factoids that he can still rattle off at will. With one of our guides, Jean-Paul Belmondo—he probably had another name, but that's who he was to me—Henry, now Henri, found a fellow atheist and self-professed anarchist who called himself a freelance philosopher. Jean-Paul had written several books, among them a humor collection for anarchists and an erotic guide to the Louvre. He had many racy tales to tell. It's hard to resist anyone who has stories about himself as a beautiful boy seduced by Jean-Paul Sartre.

At one point Henri was refusing to try *chocolat*, because it is dark and not the Nestlé's pap he was used to. And the guide said, "Henri, remember when you were about ten, and someone mentioned a girl and you said, 'Ew, yuck, ptui.' Remember that? Well, now if I said to you, 'There are four girls waiting in my apartment, and they are all twenty-two, and they all want you, and I will give you the key to my apartment'—would you take the key? I think you would. Well, *that* is how you will feel about *chocolat* in a few years, when your tastes mature." I realized this guide should also be a life coach.

Despite the fact that I managed to find ghastly food (who eats badly in France?), it was the first semisuccessful trip I'd taken since my kids were born. The only serious fight we had was about Brexit. We happened to be in Paris the night that the UK decided to withdraw from the Common Market. Because his father is British, Henry has dual citizenship. But I think he developed an opinion about the whole thing ten minutes before the count began. He could not stop watching. The count went on most of the night, and I woke up to Henry pounding his fist into the mattress. In the time it took the UK to leave the European Union he had become, in his mind, 100 percent British. "Now I will NEVER be able to work in Europe. That bloody Boris Johnson has ruined EVERYTHING."

"Can we worry about your theoretical lost opportunities in your theoretical career when it's not three a.m.?" I said. At which point he threw his pillow at the TV and started sobbing. Some of us don't do well with jet lag.

But that was our only bad moment. Flush with success, I came back home and decided to tackle the other vacation hurdle: Gus. We would have a purpose. We would go to Disney World to see villains.

Even though villains were his current obsession, his

immediate answer to going anywhere would be "No." So I knew he needed a special invitation. Gus had begun an email correspondence with Maleficent a few months earlier. He wanted to know how she stayed so evil—did the cloud of fog help?—and wondered if, since they both loved music, they could be friends despite her villainous ways. (Your child can correspond with her, too, at maleficentmanhattan@gmail.com. Also with the tooth fairy at fairyfairnyc@gmail.com. I need to get more hobbies.)

So for this trip, Maleficent needed to make a personal plea:

> *Dear Gus: I've been thinking about you. It's been a busy time here, what with the new evil spells I've had to concoct, and I've spent most of my days turning into a dragon. But I'd love to meet you one day. Why don't you and your mother drop by Disney World?*
>
> *Your friend,*
> *Maleficent*

In my haste to write this email, I actually wrote, "Why don't you and your mother drop by Israel?" so there was initially some confusion. But when I made it clear Maleficent was in Orlando, it seemed we were on our way.

When it comes to characters like Maleficent—and Siri, and a host of other fictional creatures—Gus knows they are not real. Sort of. F. Scott Fitzgerald said the test of a first-rate intelligence was the ability to hold two opposing ideas in mind at the same time and still function. This is how I choose to look at the whole situation.

When I told Gus we were visiting Maleficent, he glowed with excitement, and I got a great deal at the otherwise pricey Grand Floridian, because a small child had just been eaten by an alligator there. (A bargain is a bargain.) I promised to bring a box of Cheerios because, shockingly, the Happiest Place on Earth has a deal with Kellogg's instead of General Mills, meaning there is not a Cheerio onsite. (Autism awareness is all very well, but the real point of this book is to make Cheerios available at Disney World. Write an outraged email to wdw.public.relations@disney.com and tell them Gus sent you.)

For several nights before leaving, though, there was a lot of processing about our previous trip to Disney World a few years earlier, when the boys were ten. Henry enjoyed it mostly because with Gus we had a disability pass so we would not have to wait on the longest lines. I thought we'd reserve it for moments when Gus got really antsy, but I should have realized it was Henry who hated to wait on the lines. "AUTISTIC KID HERE, COMIN' THROUGH" he'd shout everywhere. After a day I got him to stop, but not before he tried to find out if the Autism Free Pass, as he called it, also got us discounts on the food.

That trip was before Gus's Disney villain obsession kicked in, but it fueled a fear of thunder and lightning that was already firmly in place when our monorail got hit by lightning in a flash storm. It really wasn't as bad as it sounds—Disney being famously prepared for every eventuality—though it did knock Gus and Henry off their seats, and in the quiet that followed when the power went down as we waited for instructions, Henry gleefully announced, "I think we have a lawsuit here." But Gus never quite got over it, and now we had to have many conversations about what the phrase "lightning doesn't strike twice" means. For once, I wasn't trying to make him understand an abstract concept that was beyond him. I wanted him to take to heart its literal meaning.

Of course I should have checked. As it turns out, Maleficent and almost all the other villains only hang out at Disney World for a few weeks around Halloween (I've got to get into the Villains Union). No Captain Hook, no Cruella de Vil, damn it. When I told him I had called everyone I could think of, and I was sure this information was correct, there were some tears. I considered sending Gus another note saying Maleficent really *was* in Israel (because that's what the Middle East needs—more villains), but instead I told him that she and her friends had been called away on some evil business that she had to keep secret. She was sure, though, they'd meet again.

I was in a panic. I needed villains.

Which is how we ended up one evening at the Cinderella's Happily Ever After Dinner with Cinderella, Prince Charming, and, most important, her diabolical stepmother, Lady Tremaine, and evil stepsisters, Anastasia and Drizella.

This is what happens. You're eating a buffet dinner with

Cinderella characters at a Disney resort with your *teenage son*, who is totally jazzed and rather noticeable amid the sea of five-year-old girls twirling in their princess costumes. Your choice is either to question all your life decisions or to develop Stockholm syndrome. So suddenly you're all "WHERE'S PRINCE CHARMING?" though your son is explaining to you why he doesn't really want a photo with the prince himself, because that would be weird.

The moment I stopped inwardly rolling my eyes at everything Disney was the moment Anastasia and Drizella came over to our table and absolutely thrilled Gus. They were perfectly in character, and somehow managed to convey their evilness while being kind of adorable. Gus got one of them to hiss like her evil cat.

And Prince Charming? He was probably nineteen. He called me Madam and bowed and did not seem at all perturbed that a middle-aged woman was pervily chasing him around the restaurant for photo ops. There are some things in life you can't prove but you know are true, and I know that there is a porn flick featuring young men and MILFs called *Someday My Prince Will Cum*, and Disney lawyers have not been able to quash it. Yes, they serve wine at the Happily Ever After Dinner, why do you ask?

In most of the photos, Gus looks completely insane. He left the evening babbling incoherently about villains. A bemused Disney "cast member" (restaurant hostess) had been watching him the whole evening. "Yes, he is," I said, in answer to her silent question. She is not supposed to break character, but we began talking. She used to work with kids on the spectrum, and her own son had a variety of issues, including surviving cancer when he was twelve. She explained what was up with ASD kids and the villains. The villains, she said, are painted in such broad strokes that it's wonderful knowing who they are and what they do—as compared to infinitely subtle humans, who may indeed be villains but are mostly unrecognizable as such. Disney villains (like the *Thomas the Tank Engine* trains) offer such clarity. If only we could all recognize the dastardly among us by their laugh or their eyebrows.

For Gus, it was the first time that someplace other than home truly was the Happiest Place on Earth. Maybe we'd never visit Normal Land, but we could still have fluffy bathrobes and 3,000-thread-count sheets.

Eight

DOC

The endocrinologist shows me the chart. At first she says nothing. *I hope she uses her words,* I think. I can't really read charts, but I don't want to admit this. I got a 490 on my math SATs, not that that number is seared into my brain or anything. Anyway, all I see are some lines going up and a dot way below those lines, and that dot is fourteen-year-old Gus.

"Gus was in the bottom 5 percent for height for his age throughout most of his life," Dr. Gabrielle Grinstein begins, "and now he's at the bottom 3 percent. That's not a huge drop-off, but a blood test might tell us a little more . . ."

Gus had quietly been playing *Disney Villains* on my phone, but now he is paying attention. "I have to have a *blood shot?*" he asks nervously. I don't mention that if things are the way I expect they are, that one shot is the least of it. But one thing at a time, right?

Gus is short. Not Lollipop Guild short, but close. His weight is in the twenty-fifth percentile, while his height now is in the third percentile. At fourteen, he is not yet five feet tall.

I try my best to avoid doctors. I mean, not entirely. Do you have a spear sticking out of your head? OK, fine, let's go.

Otherwise, no. The best advice my mother, a doctor, gave me was: Don't go to doctors. They will find a problem, whether or not you have one. Or, failing that, they will judge you. Particularly for fretting about something as seemingly inconsequential as height. A few years back these concerns sent worried friends with a mini son to the endocrinologist for testing. The doctor stared them up and down. The husband is five foot four and the wife, four foot ten. Finally he said, "So, what were you thinking here? That he'd be playing for the NBA?"

But this was different. John and I are not giants, but we're not wee, either. I'm five foot eight, and John, height reduced from age and now about five foot seven, swears his draft card had him at five foot ten. "Men all lie," says Dr. Grinstein cheerfully when I give her his stats. "Let's go with five feet, nine inches."

There is no surefire formula for predicting a child's height, but the estimate goes like this: add the mother's height and the father's height, add five inches for boys or subtract five inches for girls, and divide by two. That's five foot ten, in our case. A child generally falls within four inches of this height estimate, which would mean Gus should be anywhere between five foot six and six foot two. Should, but will not. Even making five foot six, which would be great, is extremely unlikely at this point.

Dr. Grinstein explained to me that the blood test she was taking may not even show a growth-hormone deficiency. To really know if he had one, Gus needed to have his blood levels checked in a hospital setting over a period of several hours. But there was another reason he might be so small. Gus was identified as "small for gestational age." He was three pounds, eleven ounces, born at thirty-three weeks. That is considered unusually small even for a twin, given the number of weeks he was cooking inside me. About 20 percent of kids who are small for their gestational age have a lifelong problem with growth hormone. It's not that they don't have it; it's that the hormone fluctuates in such a way that they never catch up. Henry was born at three pounds, one ounce, even smaller for gestational age, though now he is taller than me.

Getting Gus to give up his blood was easier than I thought; it only required three grown women shimmying and shouting "YAY" and "LOOK OVER HERE" while Gus stared at the blood coming out of his arms. He had his eye on the real prize, though, a Starbucks vanilla bean Frappuccino. It's a drink for which I give thanks daily, as it's the only bribe that works. Even that didn't work for peeing into a cup, though. We had to give up, but not before we had dropped five cups in the toilet. We went home a little exhausted, with Gus mumbling under his breath every few minutes, "I'm a brave boy." Now we wait.

"Why can't *I* get growth hormone?" Henry asks when I told him Gus might get it.

"Because you're not short?" I replied.

"Maybe I've stopped growing already," Henry continued. One of Henry's greatest talents is making himself anxious. "Maybe I'm as tall as I'll ever be, *right this second*. And you know as well as I do that extra inches of height are correlated with more success. Would you stop me from being as successful as I possibly could be?"

"It requires a shot every day for the next few years," I say.

There was a pause.

"Who wants to *loom* over other people?" Henry says. "Girls like guys who are midsize."

* * *

There are several factors militating against any sort of intervention. For one thing, I am one of the taller people in my family, with many on the Italian side resembling fireplugs. So maybe Gus just got some throwback genes. And since Gus had always been short, I had my list of extremely hot, compact men ready to hand him at the first available opportunity. Mark Wahlberg, Kevin Hart, Humphrey Bogart. Prince was five foot two. You get the idea. One of my most cherished relationships had been with a man several inches shorter and considerably scrawnier than I was; once I got over the Chihuahua–Great Dane visuals playing in my head, it was fine. It ended terribly, but every relationship ends terribly if you don't end up together, so I counted it as a great success. My takeaway from that, which I hoped to impart to my son in some maternally appropriate fashion, was this: short guys aim to please. One of the prejudices against short men is that women want to feel feminine, and culture has told them that a big guy will make her feel adorable and protected. But things are changing. Make a woman feel feminine, sure, but also powerful, and all will go well. Have you ever seen Mick Jagger with a woman who *wasn't* a head taller than he was? Shut up then.

And then, there was this: Gus had not spent one second of his life worried about his height. The parents I knew with extremely short children were generally driven by their children's fretting, not their own. So why was I dragging my perfectly content kid off to the endocrinologist to see if he was a candidate for growth hormone? A daily injection that in reality *might* give him two to four extra inches, probably no more? If I was willing to do this, what else about cosmetically improving my child seemed reasonable? My officemate, Spencer, not a big fan of my plan, kept throwing out suggestions for a rebooted version of Gus. "I know! He's got your nose—well, your old nose. Why don't you get him your new nose? Is your plastic surgeon still in business?"

* * *

"Did you hear, there's a new breed of superlice," John said with satisfaction as I prepared to discuss Gus's height issue. "They're mutants. No over-the-counter product can—"

"Speaking of tiny mutants . . . our son . . . ," I began.

For once I had John's full attention. A man who never wants any medical intervention at all wanted this one. "Since he's starting out with handicaps, we have to do everything we can to level the playing field," he said. I'd forgotten; John always felt his career would have been better if he were taller, though arguably in opera, if you're five foot eight and 350 pounds and have a glorious voice—you know, like Pavarotti—you're still hired. Brushing aside the exceptions, John insisted that in most of life Gus's height would matter. "If you have a man who's five foot five and another who's six feet and they're equally qualified, it's the six-foot man who gets the job."

But Gus's gaining a height advantage for some corporate job he would never have or to be a nightclub bouncer was not a factor. For me, it was much simpler.

Adults should be able to do exactly what they want with their own bodies. No exceptions. But in the matter of height, there is only a small window of opportunity in adolescence when growth hormone might give him these extra inches. That window would be closing soon. He doesn't care now because he still thinks like a little kid. What about when he is twenty-five? Let's imagine. Twenty-five years old, finally with the feelings of a sixteen-year-old; now he is five foot two, not happy about it, and he can't do a damn thing. If only his mother had been able to make up her mind. For some things, inaction is a form of action, putting off the decision until the decision is made.

* * *

Most of us try desperately to do right by our children. But for many parents with the "average" kid on the spectrum, the struggle is more complicated. It's one thing to make medical decisions for someone who will never be able to make these decisions for himself or make decisions that are easily reversible with no enduring consequences; it's quite another to make them when you still don't know whether or not your kid will have the understanding and will to make them himself.

This is why contemplating Gus's medical future crushes me with the weight of the responsibility. Deciding to increase someone's height is one thing: basically it's cosmetic, and even though I'm fretting, little hinges on it. No, the medical issue that really makes me hyperventilate is fertility. It's a question all parents of special needs kids wrestle with, whether they speak of it or not. What happens when you discover a lack of social skills isn't a surefire method of birth control? That the kid you think would be entirely unable to find a partner does just that, though his or her ability to understand what it takes to raise another human being is limited?

It is very hard to say this out loud. Let me try. I do not want Gus to have children.

At least I'm pretty sure that's what I want. Don't I?

If I had to decide based on the clueless boy I know today, it would be easy: Gus should not be a parent. Not just because he's still shaky on the whole concept of where babies come from, but because the solipsism that is so much at the heart of autism makes him unable to understand that someone's needs and desires could ever be separate from his own, let alone more important. He can't even quite fathom that the people he loves existed before he did. For a long time he thought I was born in 2001—his birthday. In a sense I was born then—born a mother—but I'm pretty sure that's not what he meant.

Nobody wants to visualize their child that intimately, but when I think of Gus in a sexual situation, it generally has a *Benny Hill* soundtrack. And anything with that music does not end well.

A vasectomy is so easy. A couple of snips, a couple of days of ice in your pants, and voilà. A life free of worry. Or one less worry. For me.

How do you say "I'm thinking about sterilizing my son" without sounding like a eugenicist? I start recalling all the people, outliers in some way, who had this fundamental choice in life stolen from them—sometimes cruelly, sometimes by well-meaning people like their parents. The eugenics movement can be traced back to psychiatrist Alfred Hoche and penal law expert Karl Binding, who in 1920 published a book called *The Liberation and Destruction of Life Unworthy of Life*. Its popularity fostered the first eugenics conference in the United States in 1921. The term "eugenics" means "the good birth." Sample papers: "Distribution and Increase of Negroes in the United States," "Racial Differences in Musical Ability," and "Some Notes on the Jewish Problem."

"Liberation" is such a wonderful euphemism, and in this context many people like my son—and undoubtedly some even less impaired—were "liberated" from the burden of life by those enthusiastic proponents of culling the herd, the National Socialists. An estimated four hundred thousand "imbeciles" were euthanized during Hitler's rule, but not before they were the subjects of all sorts of medical experimentation. For a while there, Austria seemed to have cornered the market on brains in jars.

The idea of outright murdering "nature's mistakes," as the disabled were called, was softened somewhat in the United States. As the psychiatrist Leo Kanner was observing and defining autism, he was also lobbying for sterilization, but not death, of disabled populations. This was considered a progressive view at the time. (He believed there were all sorts of repetitive tasks autistic people could perform that would be good for society, and he wasn't wrong here, that's for sure. But we didn't have computer programming at the time, so he proposed a population of ditch diggers and oyster shuckers.) Around the same time Hans Asperger, the Austrian pediatrician who was the first to identify autism as a unique mental condition, was concluding that "not everything that steps out of the line, and is thus 'abnormal,' must necessarily be 'inferior.'"

That was an even more radical line of thought, and one society struggles with to this day. But wherever you stand on this question, when you start considering how the history of disability is inextricably intertwined with the history of euthanizing and enforced sterilization, the mind recoils in horror. We need the DNA of neurodiverse people; some of those people have changed the world.

Inevitably, I began to question my certainty that Gus

should never have kids. There is a good and growing success rate in vasectomy reversals. But reversal is still iffy. Currently in the pharmaceutical pipelines: long-term birth control for men via hormone injection.

There are still too many side effects, but eventually they will work out the kinks, and when they do I'm going to be the first to sign him up. Kids at twenty or twenty-five? I can't imagine it. Thirty-five? I can hope.

"I'm never going to have kids, but if something happens by mistake, he can borrow mine," Henry says when he overhears me discussing with John whether or not Gus should ever have children. "He might make a good uncle. He can show them how to play piano and get around the subway systems by themselves."

Gus wanders into the room. "I like babies," he says. "They have the best feeties."

"Go ahead," said Henry, smirking. "Ask him where babies come from."

Gus changes the subject.

* * *

Dr. Grinstein, the endocrinologist, calls. "Everything is normal," she says. Gus doesn't have a quantifiable growth hormone deficiency, but the diagnosis of "low birth weight for gestational age" still stands. "The good news is that his bone age is younger than his actual age. So you have a little more time to figure out what you want to do. Come back in

November and we'll check again. Let's see if he has a growth spurt on his own."

I am nothing if not a procrastinator, so the ability to kick the can down the road, even for a few months, is a relief. It seems he may grow a bit on his own. It also seems he will *not* be in danger of dating anyone anytime soon. That decision, too, can wait.

I don't mean to visit my own worries on Gus. Yet sometimes, in moments of weakness, I do.

"Honey, do you care that the kids in your class are a lot taller than you, even the girls?"

"Nah," he says, wrapping his arms around me. "They think I'm cute. Aren't I cute, Mommy?"

You're cute.

Nine

SNORE

Henry and I are staring down at Gus, who is forming a Keith Haring silhouette on my bed, lightly snoring. "Face it, Mom," Henry says. "This is creepy."

It is not creepy. It is sweet. Or maybe it's creepy and sweet. I don't know. I just know that this has been going on for years, and I can't make it stop.

Let's recap: Only child here, no idea how babies operate. Until forty I did everything in my power to avoid being around children. Husband last spent time around children when he had one at nineteen, and now is a new father again at seventy. So our knowledge of what's normal and what's not normal is a little shaky.

When your children are born, one of them is crying, demanding, in-your-face. Gimme gimme gimme. Food now. Change now. Pick me up. Lock eyes with me so I know I exist. You know: a baby. The other one is sweet, undemanding, and limp. He never looks at you and seems perfectly content being left alone. An angel. Maybe he has other concerns. Maybe he's a deep thinker.

I never nursed Henry and Gus. Instead, when I fed them, I'd prop them up on my knees, stick bottles in their mouths, and yak or sing show tunes to them. Henry would gulp milk voraciously and look vaguely irritated that he couldn't tell me, as he did in later years, how much he hated musicals. Gus would stare off into some distant point past my shoulder, like someone at a party looking for a better person to talk to. Perhaps he was listening to some music playing in his head. But I don't know, he seemed to enjoy himself. Well, except for the part where he projectile-vomited a lot because it turned out he was lactose-intolerant. But other than that endearing little quirk—happy.

The only thing about him that bothered me was that he really didn't like to be touched. He would cry, tense up, turn his head away. But I don't particularly like to be touched, either; when I think about my Circle of Hell, it would involve eternal massage. Is it so terrible that he wasn't tactile?

Apparently it is. Apparently doctors don't like to hear, "Oh, he's just fine on his own." In *The Siege*, the classic 1967 memoir that was the first to describe raising an autistic child, the author, Clara Claiborne Park, who had three children previously, knew something was wrong because her daughter asked for nothing: "If it's all one to you whether mama comes or not, you aren't likely to call her. If you don't want teddy enough to reach for him with your own hand, you will hardly ask for him with a word." I used to be pleased with what an easy baby Gus was. He didn't ask, he didn't reach, he didn't demand. There was that touch thing, though.

When a couple of friends suggested I take him into bed with me and snuggle, I scoffed. First, snuggling wasn't in his bodily vocabulary. And second, I wasn't one of those hippie mothers. Gross. Attachment parenting, a phrase coined by the pediatrician William Sears, is based on a theory of developmental psychology that suggests a child's emotional makeup is largely determined in the first few years of life. Being constantly available and sensitive to the child's needs not only helps the child form a strong bond with the parent, but also imparts a sense that the world is a safe place. The family bed—co-sleeping with your children—is an important tenet of attachment parenting.

But couldn't I do something other than sleep in bed with him to prove the world was safe? Maybe he could sleep with a copy of my life insurance policy curled up in his little fist. The family bed struck me as one of those constructs for people who wanted to avoid intimacy with the people they were supposed to be intimate with, namely their partners. If you allowed your child to become your cuddle bunny, what other kinds of adult intimacy were you avoiding? I read a study that said that about 40 to 60 percent of Americans slept with their dogs. (The number varied with size of dog. Here, apparently, Pomeranians get lucky more often than Newfoundlands.) I would never even let my beloved golden retriever, Monty, sleep with me. So what chance did Gus have?

But OK, fine. I began doing a little reading, and maybe there was something to this bonding, even though I just thought, *Everyone is in a rush, he'll bond in his own time.* I began sticking him in my bed. On nights when John was around, I'd put Gus back in his own bed. But even though Gus was late to every party developmentally, he did, around the age of three, realize he could simply get up himself and join one or both of us.

At first a night with Gus was like having a deep-tissue massage, because some part of him would be not so much cuddling as pressing my arms and legs, again and again and again. So then I would wake up, hoist him to the other side of the bed, and have an hour's sleep before he would creep back and the kneading would begin again.

Recently I learned that the reason he did this was that, like many autistic people, he has a problem with proprioception, which is the concept of knowing where your body is in space. He didn't have a firm idea where he ended and other people began. By pressing into me over and over, he was using my body to orient himself while he slept.

Now I understand, just as I understand why he still tends to bump into people on the street. But all I knew at the time was (A) I was constantly jolted awake, and (B) in the morning I'd have bruises.

Yet soon I noticed something. After a few months of this, Gus was looking at me. Not at others, but at me, and at his father and brother. He was also not cringing when you touched him. As time went on, far from pulling back from people, he would not be able to pass those he knew without at least a fist bump. To me and his father, he was a serial hugger—so much so that I had to create the Rule of Three: you are not allowed to attach yourself to my waist, limpetlike, more than three times at any given public event. And if you hold my hand, you cannot continually kiss it as we walk down the street. Even these days, if I don't stop him, he will still do this in time to music playing in his head. I know he's thinking of the "Hallelujah Chorus" from Handel's *Messiah* when I get this on the back of my hand: *KISSS, KISS KISS KISS. KISSS, KISS KISS KISS. Kisskisskisskiss, kisskisskisskiss. Kiss KIIII-SSSS, kiss kiss . . .*

When I'd give him my lecture about the difference between public and private, that we don't behave that way in public, he would counter with, "But I just love you so much, Mommy." Try arguing with *that*.

I told myself I was lucky. And in fact I was. Up to 80 percent of all children with ASD have serious sleep disturbances. Sometimes the causes are clear—for example, epilepsy or medications that interfere with sleep. Often they are not that clear-cut. Kids who are more sensitive to sensory stimuli may be unable to filter out the street noises or, if you live in the country, the crickets, owls, or—well, anything. There is a theory, too, that involves the hormone melatonin, which normally regulates sleep-wake cycles. To make melatonin, the body needs an amino acid called tryptophan, which research has found to be either significantly higher or lower than normal in children with autism. Typically, melatonin levels rise in response to darkness (at night) and dip during the daylight hours. Some studies show that children with autism don't release melatonin at the correct times of day. Instead, they have high levels of melatonin during the daytime and lower levels at night—thus wreaking havoc with their sleep cycles.

But I was lucky because Gus could sleep well—as long as he was next to me.

Years passed. I told myself, firmly, that nine was the final year for sleeping in Mom's bed. Ten. Eleven, no more. Twelve, listen, he looks like an eight-year-old, it wasn't that bad. Thirteen, he looks nine, though with the tiniest hint of a mustache.

Henry cannot believe I am this much of a pushover. But at night I'm weak. I remind him how I was when he would come into my room late at night when he was young. I remember him at four, running into my room at three a.m., for some reason screaming, "I DON'T LIKE WHALES." At first I would hold him, tell him how intelligent and harmless they were, how important to the ecosystem. That would last thirty seconds. When he still didn't shut up, I changed tactics. "You're absolutely right, a whale is going to come into this house and eat you if you don't go back to sleep *right now*."

But Henry didn't buy my excuses. Why didn't I just lock my door? He was right. For years I couldn't bring myself to lock Gus out. Finally I did. But Gus lies in wait. A trip to the bathroom, and I would come back to a small person sprawled in my bed. Then there are the three a.m. knocks: light but relentless.

Perpetually sleep-deprived, I want only to get back into bed. And so does he. Mine.

I try reasoning: "Honey, you know neither of us gets a good night's sleep if you're in here." "That's OK, Mommy. I sleep fine with you." (Understanding the other person's point of view: still not his strong point.)

I try shame: "Gus, big boys don't do this. Do your other friends sleep in bed with their mommies?" GUS: [Silence, and a wry smile.] He admits I am right, and yet it means nothing.

On many nights when I thought he was locked out, I would open my eyes to see his eyes, dark, pellucid, and two inches away from mine. He would be smiling. Utterly benign, it was like something out of a horror movie. "*Oh my God, Gus, what are you doing?*" He never understood why I was upset. "I just like the sounds you make, Mommy." So I tried throwing money at the problem. I searched Amazon for a white noise machine that includes, along with the roaring of the tides and the trickling of rain, the sound of a middle-aged woman lightly snoring. Or *anyone* snoring. This does not exist, and thus may become my Million-Dollar Idea.

Next thought: the mattress. I got Gus's mattress eleven years ago, when he was three. Relative to my own cushy bed, it's like sleeping on horsehair. He is always resistant to anything new, but seemed open to this idea. The mattress arrived. Gus was delighted; he lay down, and I think his exact words were "Ahhhhhh."

He snuck into my bed at around four a.m.

When I lock him out, he wanders. I wake up at three a.m. to see him staring out the window, waiting for the next ambulance to drive by or mumbling weather reports to himself.

He never seems upset, and he never seems exhausted—more like a cat that can spring into action out of a deep sleep. The only times his sleep issues have anything to do with anxiety are during the summer, when there is thunder and lightning. But on those nights, he doesn't want to sleep with me for comfort. He happily sleeps in his dark, soundproofed closet. Those nights, when his closet sanctuary trumps the comfort of my bed, are the only times I can get a decent night's sleep.

When it comes to this habit that won't die, the most guileless of children is not above a certain level of duplicitousness. Three a.m., there is the knock on the door. I ignore it. More insistent. I open the door. "Mommy," Gus begins, "I have anxiety." "You do?" I ask, not having heard him use that phrase. "What are you anxious about?" Silence. "C'mon, honey." More silence. Then a frantic rush past me, a leap into bed, sleep.

Next morning: "Gus, were you really worried about something, or was that a trick?" He waggles his eyebrows. "It was a trick, Mommy."

On his fourteenth birthday, with Henry's encouragement, I told Gus that sleeping in the same bed with Mom was against the law, that if he did this a policeman would show up at our door.

This worked really well for about five days. And then for five days after that, when Henry told him that jails were filled with kids who slept in their parents' beds. He must have been thinking it through. "Are they going to take me away?" he said. I couldn't say it. "No," I admitted. "Are they going to take *you* away?" "Well, no." "OK then!" he said brightly, hopping into my bed. Henry told me I should have faked a call to the cops. "Lies only work when you really *commit* to them," he explained, alarmingly.

I have read that kids who sleep with their parents past a certain age have low self-esteem. If you ask Gus to describe himself he'll say "really nice," "friendly," "smart," and "handsome." So low self-esteem doesn't seem to be an issue. This does not stop me from worrying. In every article you read about Sante and Kenny Kimes, the infamous mother-and-son grifting/killing team, there is mention of the fact that the grown son slept in the same bed as his mother. My bed is a California king. Gus is on the far side. Still.

John, ever the soft touch, points out what's true. "He still seems like a much younger child. If you think of him as eight and not fourteen—" John begins.

"But he IS fourteen," Henry interrupts. This has become a family discussion, with the only person unconcerned about where Gus sleeps being Gus. "You can't let him do this anymore, Mom." Lately Henry has been marching into my room in the morning, hoisting a screaming Gus over his shoulder, and dumping him back into his own bed. This makes for a not-very-calm beginning to the school day.

Sometimes, when Gus would dash into my room and immediately fall asleep, I would lie awake for a while, thinking about what sleep meant, its healing power over our bodies and our minds, how sleep and dreams have been treated throughout the history of literature. I'd start to think about the story in the New Testament of Jesus on the night before the Crucifixion. He asks his disciples to spend the evening praying with him after the Last Supper, yet every one of them falls into a deep slumber. Perhaps the New Testament writers considered that the disciples' inability to remain awake represented a figurative abandonment. Was I abandoning Gus by not doing my duty toward him, helping him grow into the independent person he should be?

Other times, I would fixate on Snow White, who slept for a hundred years until awakened by a prince. Was this symbolically about her need to attain maturity and adulthood before facing the vicissitudes of life? Like my baby here. Maybe his peaceful sleep in my bed was in some way preparing him for something?

This is the part where I tell you that my son, once so averse to touch, once someone who would look right through you, is now thoroughly connected, not only to his family but to pretty much anyone who shows him love. This is entirely true. And it's also where I tell you the *alternative* fact that Gus sleeps in his own bed now, effortlessly; no locking of doors, no skirmishes at three a.m. We have both waged a long drawn-out battle, and I have triumphed.

Well.

Today, I asked Gus when he thought he would be ready to sleep in his own bed for the whole night. He thought for a minute. "I guess when I'm twenty-one," he said.

"Why twenty-one?" I ask.

"That's when there will be someone else for me to sleep with."

Ten

TO SIRI WITH LOVE

I know I'm a bad mother, but how *bad?* I wonder for the hundredth time as I watch Gus deep in conversation with Siri. Obsessed with weather formations, Gus has spent the hour parsing the difference between isolated and scattered thunderstorms—an hour where, thank God, *I* don't have to discuss them. After a while I hear this:

GUS: You're a really nice computer.
SIRI: It's nice to be appreciated.
GUS: You are always asking if you can help me. Is there anything *you* want?
SIRI: Thank you, but I have very few wants.
GUS: OK! Well, good night!
SIRI: Ah . . . it's 5:06 p.m.
GUS: Oh, sorry, I mean good-bye.
SIRI: See you later!

That Siri. She doesn't let my communications-impaired son get away with anything. Indeed, many of us always wanted an imaginary friend—and now we have one. Only she's not entirely imaginary.

This is a love letter to a machine. It's not quite the love Joaquin Phoenix felt in *Her*, the Spike Jonze film about a lonely man who has a romantic relationship with his intelligent operating system (voiced by Scarlett Johansson). But it's close. In a world where the commonly held wisdom is that technology isolates us, it's worth considering another side of the story.

* * *

It all began simply enough. I'd just read one of those ubiquitous Internet lists called "21 Things You Didn't Know Your iPhone Could Do." One of them was this: I could ask Siri, "What planes are above me right now?" and Siri would bark back, "Checking my sources." Almost instantly there would be a list of actual flights—numbers, altitudes, angles—of *planes above my head*.

I happened to be doing this when Gus was nearby, playing with his Nintendo DS. "Why would anyone need to know what planes are flying above your head?" I muttered. Gus replied without looking up: "So you know who you're waving at, Mommy." It was then that I began to suspect maybe some of the people who worked on Siri were on the spectrum, too.

(Fun fact: Dag Kittlaus, the original cofounder and CEO of Siri, is from Norway and reportedly named the app after Siri Kalvig, a beautiful Norwegian meteorologist; Kittlaus has mentioned in interviews that he is "a total weather freak.")

Gus had never noticed Siri before, but when he discovered there was someone who would not just find information on his various obsessions—trains, buses, escalators, and, of course, anything related to weather—but actually semidiscuss these subjects tirelessly, he was hooked. And I was grateful. Now, when I would rather stick forks in my eyes than have another conversation about the chance of tornadoes in Kansas City, Missouri, I could reply brightly, "Hey! Why don't you ask Siri?" And not only would Siri happily give him tornado reports for the entire Midwest, but upon being thanked she'd chirp back, "I live to serve."

It's not that Gus believes Siri's human. He understands she's not—intellectually. But like many autistic people I know, Gus feels inanimate objects, while maybe not possessing souls, are worthy of our consideration. I realized this when he was eight, and I got him an iPod for his birthday. He listened to it only at home—with one exception. It always came with us on our visits to the Apple Store. Finally I asked why. "So it can visit its friends," he said.

So how much worthier of his care and affection is Siri, with her soothing voice, charm, helpfulness, puckish humor, and capacity for talking about whatever Gus's current obsession is for hour after hour after bleeding hour?

Online critics of personal assistants have claimed that Siri's voice recognition is not as accurate as the assistant in, say, the Android, but for some of us, this is a feature, not a bug. Gus speaks like he has marbles in his mouth, but if he wants to get the right response from Siri, he must enunciate clearly. (So do I. Since I'm the one with an iPhone, I had to ask Siri to stop referring to the user as Judith and instead use the name Gus. "You want me to call you Goddess?" Siri replied. *Why yes, and could you make your voice sound like Alan Rickman's?*) Also wonderful for someone who doesn't pick up on social cues: Siri's responses are not entirely predictable, but they are predictably kind—even when Gus is brusque. I heard him talking to Siri about music, and Siri offered some suggestions. "I don't like that kind of music," Gus snapped. "You're certainly entitled to your opinion," Siri replied. Siri's politeness reminded Gus what he owed Siri. "Thank you for that music, though," Gus said. "You don't need to thank me," Siri replied. "Oh yes," Gus added emphatically, "I *do*." Siri even encourages polite language. When Henry egged Gus on to spew a few choice expletives at Siri, she sniffed, "Now now. I'll pretend I didn't hear that."

I was very curious about Siri, which is how I found myself having martinis with Bill Stasior, who bills himself on his website as the VP of Siri, a husband and dad, and a pug owner. He is the least intimidating genius I've ever met. He has devoted much of his life to machine intelligence, first at MIT, then Amazon, then Apple. When he got to Apple, Siri was considered, he says, a "trouble child." "It was bad at understanding what you said, and a lot of times wouldn't answer you at all. As more and more people used it, it went from kind of a cool demo to a disaster," Stasior said. Siri had too many misfires to count; if you searched for "sadness," for example, Siri came up with the stadium for the Cleveland Browns and a predictable uproar from Cleveland Browns fans ensued. (The Browns are a notoriously losing team, and a comic's viral video that called their stadium "The Factory of Sadness" gave Siri this verbal association.) Siri got more and more clever, but new issues continued to crop up. There was a bit of a furor when it was discovered that if you asked Siri whether doctors were male or female, she said "male," a mix-up having to do with the words "male" and "mail." Now if you ask, Siri has gone all Rosie the Riveter on us. "In my realm," she says, "anyone can be anything."

There is a great deal of thought—and manpower—that now goes into Siri's politeness. "We call them our conversational interaction engineers," my new pal said. "We've really given a lot of thought to who Siri is. Have you ever read *The Hitchhiker's Guide to the Galaxy*? There's this character at the beginning of the book, an alien who is trying to pretend he's a human. He picks the name Ford Prefect. It's just odd, it's not wrong. So Siri's like that—a little nerdy, not quite savvy enough to be cool. Funny, but a little off." And every now and then she is allowed to be a bit snotty. Go ahead, ask Siri the unanswerable mathematical question: What is zero divided by zero? This is how she answers:

IMAGINE THAT YOU HAVE ZERO COOKIES AND
YOU SPLIT THEM EVENLY AMONG ZERO FRIENDS.
HOW MANY COOKIES DOES EACH PERSON GET?
SEE? IT DOESN'T MAKE SENSE. AND COOKIE
MONSTER IS SAD THAT THERE ARE NO COOKIES,
AND YOU ARE SAD YOU HAVE NO FRIENDS.

That's definitely edgy for Siri. But there is thought—a lot of thought—that goes into how Siri treats people who might be seeking information when they're really upset. It's still a work in progress, though. If you say, "I have been raped," Siri brings up the number of a national rape crisis hotline. Ditto the National Suicide Prevention Lifeline if you say, "I want to kill myself." But if you tell Siri, "I feel like killing my husband," she says either "I don't know how to respond to that"—or she tries to find a movie called *Killing My Husband*.

* * *

Of course, most of us simply use our phone's personal assistants as an easy way to access information. For example, thanks to Henry and a question he asked Siri, I am now well-acquainted with a celebrity website called herbrasize.com.

But the companionability of Siri is not limited to the communications-challenged. We've all found ourselves having little conversations with her (or him—you can make the voice male) at one time or another. Occasionally I see this in restaurants: people seemingly not insane talking to Siri on their phones or iPads.

"Siri and I have a very strained relationship," says one of my friends, the writer Nancy Jo Sales. "She's very passive-aggressive with me and once I told her so. She said, 'Nancy, I'm doing the best I can.'"

"I was in the middle of a breakup and the guy was AWOL, and I was feeling a little sorry for myself," says another pal, Emily Listfield. "It was midnight and I was noodling around on my iPhone and I asked Siri, 'Should I call Richard?' Like this app is a Ouija board. Guess what: *not a Ouija board*. The next thing I hear is 'Calling Richard!' and *dialing*. At which point I realized I was screwed. My daughter assured me there's a two-second rule—the call doesn't register if you hang up really fast—but I knew she was lying for my benefit." Listfield has forgiven her Siri, and has recently considered changing her into a male voice. "But I'm worried he won't answer when I ask a question. He'll just pretend he doesn't hear."

Siri can be oddly comforting, as well as chummy. "I was having a bad day and jokingly turned to Siri and said, 'I love you,' just to see what would happen and she answered, 'You are the wind beneath my wings,'" says one friend. "And you know, it kind of cheered me up."

(Of course, I don't know what my friend is talking about. Because I wouldn't be at all cheered if I happened to ask Siri, in a low moment, "Am I ready for a face-lift?" and Siri answered, "You look fabulous." That wouldn't make any difference to me. Nope.)

For most of us Siri is merely a momentary diversion. But for some it's more. My son's practice conversation with Siri is translating into more facility with actual humans. Recently I had the longest conversation with him that I've ever had. Admittedly, it was about different species of turtles, and whether I preferred the red-eared slider to the diamondback terrapin. This might not have been my choice of topic, but it was back and forth, and it followed a logical trajectory, and I can promise you that for most of my beautiful son's years of existence that has not been the case.

Developers of intelligent assistants have already recognized their uses to those with speech and communications problems—and some are thinking of new ways the assistants can help. According to the folks at SRI Technologies, the research and development company where Siri began before Apple bought the technology, the next generation of virtual assistant will not just retrieve and discuss information—it will be able to carry on more complex conversations about a person's area of interest. "Your son will be able to proactively get information about whatever he's interested in without asking for it, because the assistant will anticipate what he likes," said Bill Mark, VP of information computing sciences at SRI who leads a team of researchers in developing this technology. Mark said that he also envisions assistants whose help is not only verbal but also visual. "For example, the assistant would be able to track eye movements, and help the autistic speaker learn to look you in the eye when talking," he said. "See, that's the wonderful thing about technology being able to help with some of these behaviors," he adds. "Getting results requires a lot of repetition. Humans are not patient. Machines are very, very patient."

In fact, there is a new generation of virtual assistants specifically for children. By the time you read this Mattel will have introduced Aristotle, an AI personal assistant that you place in your kid's room that is capable not only of playing favorite bedtime stories and music, but also of recognizing and adapting to young children's voices in a way that standard assistants do not. Yes, there may be a danger here of outsourcing those tender bedtime moments to a robot. I know it sounds bad. But let's get real here. How many of us would pay $300 for someone else to read *Goodnight Moon* for the four thousandth time?

And there are assistants that go a step further. One in development takes a particularly interesting approach to kids on the spectrum. The hope is to use the kid's obsession to expand his or her world. Only let me stop calling them "obsessions" or "perseverations," the terms commonly used to describe the circumscribed interests of the person with ASD. Let's call them "affinities." That's what the creator of this app, Ron Suskind, calls them. Suskind, whose book *Life, Animated*, and the Oscar-nominated documentary it inspired, chronicles how his autistic son, Owen, emerged from a nonverbal world via engagement with Disney characters, hates the pejorative term for a subject a person loves deeply. Is an obsession just an annoying nuisance? He doesn't think so. Suskind believes an autistic person's affinities can be "a pathway, not a prison."

Suskind's app, Sidekicks, developed by the Affinity Project, works like this: Your child has the app on his phone, and so do you. He will click on it, and a subject of his interest will pop up; maybe it'll be all about Star Wars movies or, you know, turtles. Your child can request a slice of a favorite book or movie or song, and a little avatar will pop up—the sidekick—and ask your child questions about it. What was the dragon thinking in this clip? Was he happy or sad? What does he want? Where are these turtles born? Etc., etc. Correct information will be programmed in, but there is also a "man behind the curtain"—an actual human answering the questions and engaging the child. The human—either the parent or, as time goes on, coaches who are hired to work with the app (speech therapists, psychologists)—answers the child's questions, and questions and answers are recorded and accumulate over time. But the humans are typing or speaking answers, and the answer comes out in the voice and personality of the computer avatar/buddy—in the way that Suskind and his wife talked to Owen as Disney characters before he was ready to have regular conversations with them. As odd (and time-intensive) as it sounds, the sidekick plays to many autistic kids' comfort with machines over human beings. Yes, you are again outsourcing some engagement with your child. But many of us need to, if we value our sanity. As Suskind's wife, Cornelia, told him at one point, after hundreds of *Dumbo* viewings, "If I have to watch that movie one more time, *I'm* going to run away and join the circus."

Right now Sidekicks is still a pilot program. There is a waiting list of several thousand parents who want to test it. Companies responsible for Siri, Alexa, and other virtual assistants are aware of Sidekicks, and there are efforts to see how their technologies can be incorporated. Suskind hopes Sidekicks will be ready for more widespread use (Suskind sees it as a subscription service, maybe twenty bucks a month) by 2018, and hopes professionals who work with the ASD population will want to learn how to use it to work with their clients. I kind of hope that the avatars will expand to be specific characters Gus will want to connect with—a subway train, maybe, or Lady Gaga. (Frankly, who *wouldn't* want to have Lady Gaga explain life to you?) But regardless of that, there is something deeply cheering about the idea of Sidekicks. A buddy even more customizable than Siri who will eventually get to know the likes and dislikes of his particular Hero intimately. That's what Suskind calls the kid/end user: the Hero. Why? "It was my son Owen's idea," Suskind tells me. "It's his understanding of Disney characters and their sidekicks: 'A sidekick helps a hero fulfill his destiny.'"

I asked Mark whether he knew if any of the people who worked on Siri's language development at Apple were on the spectrum. "Well, of course I don't know for certain," he said thoughtfully. "But, when you think about it, you've just described half of Silicon Valley."

It is a slow process, but I am accepting that what gives my guy happiness are not necessarily the same things that give me happiness. Right now, at a time when humans can be a little overwhelming even for the average kid, Siri makes Gus happy. She is his sidekick. One night as he was going to bed there was this matter-of-fact exchange:

GUS: Siri, will you marry me?
SIRI: I'm not the marrying kind.

GUS: I mean, not now. I'm a kid. I mean when I'm grown up.
SIRI: My end-user agreement does not include marriage.
GUS: Oh, OK.

Gus didn't sound too disappointed. This was useful information to have—and for me, too, since it was the first time I knew that he actually thought about marriage. He turned over to go to sleep:

GUS: Good night, Siri, will you sleep well tonight?
SIRI: I don't need much sleep, but it's nice of you to ask.

Very nice.

Eleven

WORK IT

"Do you know any prostitutes?" Henry asks.

Over time I've learned that what Henry asks and what I think he's asking are not necessarily the same thing. So I try to think before I answer. This time, I figure, he's asking about social classes, so I answer carefully. I tell him I do know several former prostitutes, and I launch into my lecture about how people sometimes do jobs at one point in their lives they wouldn't do at another, and while the work can be awful and exploitive, it can also be rewarding to some women. You can't make any assumptions about intelligence or morality. "The preferred term is 'sex worker,'" I add primly.

"OK," he says. "But do you know any *male* prostitutes? And can they make a living without being gay?"

I then realize we're not talking about morality or class or race or politics. We're at Career Day.

"You cannot make a good living as a straight male prostitute, and even if you could this wouldn't be a good choice for you," I said.

"You always told me I could be anything I wanted to be," he said, a little sulkily.

I guess I had been talking a little too much about finding a future career. But I couldn't help it. Freud said the two most important things in life are work and love, and I couldn't agree more. I have been working since I was twelve, when my parents got me a newspaper route around our suburban neighborhood. Being unathletic, I couldn't balance a bike with a huge basketful of papers, so my mother slowly trailed after me every morning in her car, holding on to most of my stash of papers and thus allowing me to do my job. While I never wanted to babysit, because I didn't like or understand children, at one point I figured I had to. I lied about my age by a few years and nobody checked. I ended up babysitting for a precocious boy who was about six months older than I was. He was on to me. This made for awkward bedtimes.

Mostly, though, my parents got me dog-sitting jobs. We didn't really ask too many questions beforehand about the dogs I cared for, which is why I'm lucky I was never maimed. I still remember a springer spaniel named Bella: once she'd eaten, she'd retreat to the back of my mother's closet. Safely ensconced in fake fur and polyester, she'd growl ominously if anyone approached the closet. My mother had to wait until Bella was out of the closet before she could get dressed. Sometimes she could entice her out with a tennis ball. We called this tactic "Bella of the Ball" because we were such geniuses.

By the time I got to high school, I stopped relying on my parents to get me jobs. These days, high schoolers burnish their college applications with exotic internships. But this was before privileged teenagers could be sent to observe meerkat fields in the Kalahari Desert, so I was selling midprice handbags in a mall. I loved this job. I started out every day with a sinister thought: *What is the ugliest bag in this store, and can I sell it?* This goal would consume my day. But after a few weeks of this, I began feeling sorry for the bags, sorry that I had thought of them that way. So instead of looking for women I didn't like to buy my ugly bags—say, the Dallas handbag inexplicably shaped like a telephone, complete with cutting-edge push buttons instead of a rotary dial—I began targeting Purse Angels, nice people who would give my Charlie Brown leather goods a forever home. I began to take it very personally when women turned up their noses at whatever I praised. Fortunately I only did this job for one school year, and never revealed my state of mind to the woman who owned the store. But being emotionally invested in cheap leather goods made me realize that perhaps sales was not my calling.

But no matter, I liked to work. I couldn't imagine *not* working. A solitary person, but one who was nevertheless deeply nosey, I loved how a job enabled me to engage with people and ask questions that would have been impolite otherwise. *What occasion do you need this purse for? Your son's wedding? Congratulations! How do you feel about the bride? You don't like her? TELL ME MORE.*

Since many of my fondest memories of high school involved not hanging out with friends on weekends but clocking into some job, I couldn't help but feel that that was the key to happiness for my own kids. Sometime. In the future. Or maybe now, at least for Henry? I reminded him, perhaps a bit too frequently, that he is someone who likes material possessions. I showed him his Christmas list versus Gus's list that I had saved from when they were ten years old:

> Henry: *Club penguin cards; alot of stuff; play mobile; froot loops; presents; Pez; 100 buks, pokemon cards; morio and sonik at the Olimpic winter games, a DS, an iPhone, ds games, a Scotland uniform, I pod touch.*
>
> Gus: *I want Daddy to come home.*

"So, you're the kind of person who needs a lot of money."

"Mom, have you noticed that I'm fourteen years old? Who is going to hire me to do anything at fourteen?" he said. When he started in with this entirely reasonable argument, I pulled out my Matthew Freud story. Matthew Freud—great-grandson of Sigmund, former husband of Rupert Murdoch's daughter Elisabeth, notorious bad boy—runs one of the largest PR companies in the UK. Years ago I interviewed him, and he told me that his first job was at nine, selling mice to kids at a local school event. Then, when infuriated parents came back to him with their child's "pet," he would take the mice back—if the parents paid him. Now *that's* entrepreneurship.

"You want me to sell mice? Now I'm confused," Henry said.

"It's not the mice. It's the gumption. Matthew Freud was only nine years old. You are fourteen. We can think of something for you."

Henry has all the tools necessary to work right now. And he does make money. Unfortunately, he makes it playing poker and taking sucker bets from his friends. One day he came home with $150 and a smile on his face. It was some football bet. He explained its intricacies, and while I didn't understand it entirely, I did understand that somehow, even if Henry were to lose, his friend Joey would owe him fifty dollars. "See, I'm selling him my football draft pick, and at fifty dollars, that's a deep discount," Henry explained.

My Nathan Detroit plays in two fantasy football leagues. One is a bunch of kids from his high school. The other is a group of attorneys from Goldman Sachs. He was invited through a family friend. I tried not to learn anything about it beyond that. But my point is, if he's not kneecapped before he's eighteen, he will eventually find a job and be OK.

And then there is Gus. Gus, whose interests and skills are limited. Gus, who is still a little unclear about what is real and what is make-believe, who thinks that everyone is his friend, who has no idea about sarcasm or competition or envy or ambition. Or the value of money.

* * *

"Bye, I'm going to work!" Gus says after dinner, as he has for the last three years. And work he does, for as long as the doorman who's on that night will let him. I don't know when he got it into his head to be a doorman, but once that idea was there, it stuck.

At first it must have been rather startling, this tiny kid

who would don the doorman's jacket and sit at the front desk. But now everyone knows him, and Gus takes the job seriously. He knows the name of everyone in the building, their dogs, their apartment numbers. He knows all the food deliverymen. The moment a person comes into the building, Gus checks the computer to see if they have a package, lets them know, and gets it from the mailroom if they do. No amount of insistence on my part can make him understand that it's rude to ask people where they are going or what they are doing that night, or who this "new person" they've come in with is—a particular problem for one man in the building who is known for having a parade of paid companions. Gus stops all the deliverymen, including the guy who delivers weed to half the building; I believe he's told Gus he works for Grubhub.

"Don't worry, we'll get him in the union eventually," says Jen, my most loving and lovely neighbor. This is unlikely. If Gus saw someone come into the building with a gun, he'd probably ask the guy what kind of gun it was and what street he bought it on. Gus can do every part of a doorman's job except the part that involves keeping people out. He'd undoubtedly welcome Charles Manson with a smile and a wave.

* * *

"Job." J-O-B. That word has music and beauty to me. It is not just about making money. It is knowing your kid will have a place in the world. Before having Gus, I read Studs

Terkel's wonderful 1974 book *Working*, in which he interviewed dozens of workers in a variety of industries. The idea that resonated most came from an editor he interviewed: "Most of us are looking for a calling, not a job. Most of us have jobs that are too small for our spirit. Jobs are not big enough for people."

Sure. I get it. But I'm betting Terkel didn't talk to any autistic people. Because he might have seen that to some people, a job is to the spirit what helium is to a balloon.

* * *

Over five hundred thousand people with autism will become adults over the next decade, and most of them won't have a job, according to a 2015 study by the A. J. Drexel Autism Institute. Two-thirds of autistic kids have no plans for either a job or further education after high school. As they get into their twenties, about 58 percent of young ASD adults have some employment, as compared to 74 percent of those with intellectual disabilities and 91 percent of those with speech impairment or emotional disabilities.

That's a lot of people with nothing to do and nowhere to go. And while of course there are those with medical and/or cognitive issues that make employment out of the question, there are many more who are perfectly employable, with a bit of flexibility and attitude adjustment. Employment that is not about charity, but about recognizing some of the nutty talents that often present themselves with autism and exploiting the hell out of them.

I am not talking about the very small percentage of exceptional autistic geek talents so well represented in Silicon Valley; the Temple Grandins and John Elder Robisons will take care of themselves. (Though even in these cases, I don't want to be cavalier. High IQ and specialized skills aren't always enough. A woman at NYU's then-named Asperger Institute told me that a substantial percentage of even her most brilliant patients couldn't stay employed because of their social cluelessness. Getting a medical degree was one thing. Practicing as a doctor, with all the people skills that involves, was quite another.) Rather, I mean the more pedestrian but still very needed tasks that play upon the autistic individual's love of repetition, or passion for categorizing objects. How many people in this world actually love to take apart electronics and sort out the parts? If you're on the spectrum, that activity might make your day. Certainly that was Bill Morris's thinking when he started Blue Star Recyclers in Denver, a company that is both cutting down on electronic waste *and* employing people who are excellent at dismantling and sorting.

Specialisterne USA originally began in Finland, when Thorkil Sonne refused to accept the idea that his autistic son, who could reproduce timetables and maps from memory, was unemployable. Today his company headhunts software testers and data entry people, jobs where the ability to perform tasks that would seem tedious and repetitive are in the wheelhouse of many autistic people. When Jonah Zimiles, an attorney and recent Columbia MBA grad, and his wife realized there were few businesses where they could imagine their young autistic son getting the kind of job he would need someday, he opened [words] in Maplewood, New Jersey—a bookstore that hires people on the spectrum. Jonah believes in "job crafting"—adjusting jobs to the individual. While there will always be some challenges, it's not that hard to find people with autism who like to categorize and order books or enter inventory on a computer, even if they may not be the best salespeople in the world.

The project that really got me excited—possibly because it will be ready about the time Gus will actually be looking for a job—is taking shape right now at Rutgers University in New Jersey. Rutgers has more than fifty thousand students spread over many campuses. It also is in the state with the highest autism rate in the country: one in forty-five kids, with a rate of one in twenty-eight boys. (Exactly why is a mystery. It's not just because people with autistic kids moved here for the solid medical and school resources, as New Jersey sources report. According to the CDC, 83 percent of the people with autism in New Jersey were born there.) Rutgers is now developing the Rutgers Center for Adult Autism Services. The idea is that about a hundred adults with autism will work in various jobs on campus, and a percentage of them will live—permanently—in housing with graduate students, who will oversee whatever the residents can't do on their own. If this were Gus, I'd guess they would be helping him pay bills and cut his meat, but by the time he's twenty-five, who knows?

"Rutgers has its own bus system to all the different campuses, so residents can learn to take the buses and get around easily," says Dina Karmazin Elkins, who, along with her father, former CBS CEO Mel Karmazin, and one other family, seeded the project for $1.5 million and is currently raising additional funds.

Colleges are a big community with various jobs that suit various kinds of people. One small example: Rutgers has its own midnight movie theater. "A number of people on the spectrum have sleep cycles that are different than ours, and may function best working at night, so some people may be well suited for working there," Karmazin Elkins tells me. Dina's excitement about the project is very personal. She currently has her fourteen-year-old autistic son working three part-time jobs.

Dina envisions work-live programs on college campuses across the country, because colleges don't tend to go out of business. She knows that the most vulnerable are the last hired and the first fired—so the idea of having programs like hers on campus is that "those jobs won't go away if the economy tanks."

* * *

After three years, Gus was fired from his job as a doorman.

I was shocked. I'd had people I barely knew stop me in the elevator and tell me how having him there to greet them so enthusiastically cheered them at the end of the day. Becky, a recent divorcée who'd been going through a rather painful time, told me Gus had gotten into the habit of waiting for her when she took her pit bull, Francesca, for their last walk of the day, then gallantly escorting them to their door. On the rare occasions when he wasn't waiting for her, she said, her day felt incomplete.

Of course I didn't hear from the people who found him annoying. And someone—or maybe several—did. Maybe they didn't like being asked where they were going by the nosey parker, or maybe there was something about having that little garbled voice buzz them on the intercom when food was being delivered that reminded them they were paying for actual union doormen, and this was a little unprofessional. Apparently, other kids in the building were asking why *they* couldn't work the door. Whatever the reason, Gus's job was no more.

I took for granted that people would make allowances for my son, that they should, because he is such a good boy. And when they didn't, and I had to tell him he was laid off, I made up the excuse that the doormen's union wouldn't allow him to work until he was eighteen. He pouted, but he accepted it. Then I went to my bedroom and locked the door and sobbed deep heaving sobs. I was embarrassed. Embarrassed that I had believed my autistic son was actually performing a service, when in fact he'd been merely tolerated, a nuisance. Embarrassed that I had the audacity to convince myself that he was actually in some sort of training, that someday he would have a job like this, that he would be just another dude with a job, a guy who'd get a million hellos.

The false job had given me false hope.

Then Gus got a little older, less pushy, and wasn't insisting on operating the intercom to announce visitors. He'd just hang out, greet people, and find their packages and dry cleaning. In this he was actually useful—or at least that's what the new doorman tells me. And I guess I need to believe it. Now, most nights, he is back with two of his beloved door guys, Jimmy and Jerry. His evening concludes when he walks Becky and her pit bull, Francesca, back to their apartment.

* * *

In truth, I don't know what Gus will be able to do. I do know that he does practice a kind of learned helplessness; I did not know, for example, that he could pour a glass of milk for himself until one day recently I got vertigo and couldn't move without being wildly nauseated, and no one was around, and Gus really, really wanted milk. It was that day when I thought about something John Elder Robison had written in his book *Switched On*, about the low expectations we have of people with autism, how it extends to everything in their lives. I flash back to how John used to carry Gus on his back everywhere until he was seven or eight; then, for a while after that, Henry carried him. Why? Because Gus liked it.

"There is an unintended downside to the new diagnostic awareness," Robison writes. "Maybe today's autistic kids were more like wise and wily pets who had trained their parents to feed them, house them and provide entertainment and healthcare for a lifetime, all for free." I love this idea, even as I know that it's largely untrue.

What upsets me almost as much as Gus not finding a job is Gus finding a job that is nothing more than charity. Kindness is fine. The pity that is at the heart of charity makes me ill—even if, quite possibly, Gus wouldn't know the difference. And then there are the real little jobs he wants to do, but he can be so easily cheated. A neighbor with a lucrative job asked Gus to cat-sit for ten dollars a day, going twice a day to feed and play with the cats. He couldn't open their food cans himself, so this was a Mom-and-Gus project. But no one was more diligent about making sure the cats were cared for. I might forget an afternoon play session; Gus never would. But I noticed that if Rebecca came back earlier than expected—say, in the afternoon rather than late at night or the next day—she would only pay Gus five dollars instead of ten because he was supposed to come see the cats twice.

Gus didn't care at all. I was so angry the third time she did this that I stopped him from cat-sitting and never talked to her again. I want him to understand money, and understand that he was being shorted. Yet I couldn't confront her. I was embarrassed because to her I would be making a stink over a few dollars. How can I expect him to advocate for himself when I can't do it for him?

I loved that it was a real job. I loved that Gus loved it. And while I had to hold Henry back from confronting her—something he would not do for himself, but would do for his brother—I loved telling Henry what had happened, for the sheer pleasure of watching his righteous indignation. It reminded me of a day we were at McDonald's when Henry and Gus were six. Henry noticed they had only put nine McNuggets in Gus's ten-pack. No amount of talk on my part could convince him not to march up to the girl at the counter and complain. Of course, I wasn't quite sure if Henry was being protective of Gus or thinking, *Hey,* I'm *the only one allowed to take advantage of him like that.*

* * *

Last year there was a viral YouTube video called "Dancing Barista." A kid with autism named Sam is a barista at Starbucks; his manager posted the video. Being a barista was the kid's dream job. But Sam had been told he was unemployable: his movements are jerky, and he can't really sit still. As he explained when he and the store manager appeared on *Ellen*, he needs to keep moving. "I can concentrate when I dance," he says.

And so he does. The video makes me cry every time, but it shouldn't, because there is nothing sad about it. The lovely Starbucks manager saw that he could make someone's dream come true. He looked past the jerky movements and the lack of conversation and saw a goofy teenager with a burning enthusiasm and a talent for making a perfect head of foam. All he had to do was let him dance.

Twelve

CHUMS

This was the dream. I had a dog—or a cat, or hamster, or snake. It lived with me, but I forgot it was there. Forgot food, forgot water. As it was withering away, I would finally notice. By the time I remembered, it was too late: I would try to feed it, I would apologize, and it would die in front of me. I had this same dream, about once a month, for years.

Then I had children. OK, maybe I could take care of living things after all.

That dream went away, but this one took its place: I am gone, and Gus lives alone. There are no visitors. Somehow—it's a dream; one doesn't dwell on the logistics—take-out food is delivered to his house. But he doesn't know how to open the container. He just stares at it, like a dog stares at a tin can.

So many times I have woken up from this dream, shaking. Will he ever learn to use a can opener? But more to the point: If he doesn't, will he have a friend who'll open the can for him?

* * *

Those are dreams. Here's real life.

"Hello."

"Hello."

"Hello."

"What are you doing?"

"I'm getting ready for school. What are you doing?"

"I'm talking to you on my phone."

"Cool."

I admit these conversations would not make Noël Coward envious. But they are conversations. Every morning for the past six months, Gus and Mandy text each other.

"Who is Mandy?" I ask one day.

"She's a friend," Gus replies.

"Where do you know her from?"

"I'm not sure."

What? Then I remember who I'm talking to. "Have you ever actually met Mandy?" I ask.

"No," he says, adding quickly, "but she's my friend."

Gus's friendships remind me of special ed report cards. In the ongoing effort not to be all judgy, report cards at special ed schools have a language all their own. When a kid has a skill set that is almost nonexistent, the report card refers to that skill as "emerging." In our house, this has become shorthand for "You have no idea what you're doing." As in, Henry, seeing me late for an appointment and frantically fussing with my phone: "You still don't know how to call a car with Uber, do you?" Me: "That skill is emerging."

So I like to say: Gus's concept of friendship is emerging.

Mandy, as it turns out, is a girl who went to LearningSpring, Gus's last school. She is considerably older than he is, so I have no idea how they ever met or how they ended up swapping phone numbers. But she is very nice and is in a job-training program. I know this because one day I swiped Gus's phone and texted her. She then asked for my phone number and email, and I ignored her request because I know better. Perhaps because many of the kids Gus knows came to this whole communication thing late, phones still have novelty value. This means that if I gave Mandy my phone number, she would text me every morning, too. So it would be me, Gus, and the other fifty people she probably has on her check-in list.

I already have this little dilemma with another kid in Gus's school. As exhausting as Gus's desire for repetition can be, he is perfectly capable of being exhausted by other people, even people he loves—and when he is, he tends to hand the phone over to me. This is how I inadvertently became friends with Aidan, a charming boy who liked to call, email, and FaceTime even more than Gus. Aidan wants to be a talk show host in the style of Anderson Cooper, and is perfecting his technique on me. After he barrages me with questions (ignoring the answers) and tells me about his new school, he'll sometimes say, "We'll be right back after this message." I've learned to wait for the commercial break he programs into his talks, and then we resume. One time he wanted to come over to our house, not to hang out with Gus but to interview me on tape. Before the interview, he was very exacting about what I should wear and how I should do my hair. The look was somewhere between Christiane Amanpour and a hooker. I pray he's lost that video.

Many kids who could barely talk in their early years develop a passion for communication later in life, especially when technology is involved. Part of the fun of listening to them is that, unlike neurotypical kids their age, they do not place any value on "coolness"; there is no moderation in their joy. So a recent conversation between Gus and his friend Ben on FaceTime went like this:

GUS: How was your day?
BEN: Fantastic! How was yours?
GUS: The best! I had a great apple for lunch.
BEN: I love apples! Mine was terrific, too!

And on and on like that, for an hour.

Partially this preference for gadgets over live face-to-face chat may be because people with autism often don't want to look at someone when they talk. Recently a rather brilliant young man with ASD told me that for him it has to do with his field of vision—he sees only a small fragment of a person when he looks at him, and his imagination fills in the rest of the visual field with something ghastly. So what he sees is not the person or people, but part reality, part phantasm. And thus just about any form of communication is easier than face-to-face. He is unusually articulate, and reminds me that there's more information like this available in his 1,200-page self-published book.

So when Gus and his cohorts drive me a little batty, I have

to remind myself: now there are connections where once there were none.

* * *

Henry's friendships are complicated, multifaceted, nerdishly competitive; there is a small tight group, and their bonding involves endless teasing and one-upping; the guy who knows the most about a particular subject wins. This was the conversation I heard a few days ago:

JULIAN: Have you watched *Vampire Diaries*? It's great.
HENRY: It's just *Twilight* repackaged.
JULIAN: It's not *Twilight* at all.
HENRY: Are there tan werewolves with six-packs?
JULIAN: Well, yeah . . .
HENRY: Are there super-pale vampires that sparkle in the sunlight?
JULIAN: Well, these vampires don't sparkle . . .
HENRY: Is the vampire family fighting with each other?
JULIAN: Um . . .
HENRY: And are the werewolves and vampires all fighting over a pretty girl with no personality whatsoever?
JULIAN: FINE.

Gus's whole friendship thing is simple: if you are in Gus's orbit, you are his friend. The deliveryman from the Chinese restaurant, the kid at the next desk in school, the train conductor who hands him the microphone, the computer he uses to see other friends, the dog walker he passes in the lobby, the caretaker he cajoles into taking him to bus stops, all of them. He doesn't have a lot of requirements to be in his legion of friends, and I'm not aware there's any hierarchy. He won't want to spend any face time with you, but that doesn't make you less of a friend; he doesn't really want to spend face time with *anyone*, immediate family possibly excepted. There is, however, a special class of friends—the people I pay to take care of him.

(Almost every woman today struggles with balancing work and motherhood; I am no exception. The difference is that I don't have the luxury of worrying about balance. I'm not a contributor to the family coffers; I *am* the family coffers. John is long retired, but not a Mr. Mom kind of dude. In short, it's convenient for everyone that I love my work since I have no choice about whether or not to do it. On a good day, I'll make more than it costs to pay the caretaker who allows me to work. As I write this, Michelle is taking Gus to Grand Central Terminal . . . so I hope you've bought this book rather than borrowing a friend's copy.)

Gus's first caretaker was Orma, a stolid, serious, deeply kind woman from Jamaica, who was with him from the ages of zero to ten. Among the many things he learned from her was that Halloween is the holiday of the Devil and it is always much colder than you think, so carry a sweater in July. Orma has never entirely accepted that she doesn't work for us anymore, so she still drops by with her newest little charge, raids the fridge for a Diet Coke, and, refreshed, takes off. One time, on a very hot summer day not that long ago, I came home to find her taking a shower. She's very comfortable in my home.

She adored John, and didn't bother to hide her belief that he was the better parent, perhaps because her expectations of men are pretty low. John got a trophy just for showing up. I, on the other hand, had a great deal to learn, and she tried to teach me every day for ten years. We had the kind of resentments that build up over a decade, the divides of race and class where the employer feels judged and guilty, and the employee knows it and smiles to herself. But Gus loves her, still hops when she comes by, and is not in the least perturbed when she invites herself to his birthday parties. Orma and I put aside our quiet mutual resentments during the most recent presidential election where our worries, different as they were, brought us closer. I buy a lot of extra Diet Coke, and it will always be waiting for her.

Kelly was a twenty-five-year-old hipster who was getting her teaching degree, and had to live through the shouting of Henry, then ten, who thought homework was for suckers (and still does, but at least he doesn't shout about it). She was fun and smart and much more interested in her friends and family than her job, which suited me wonderfully after Orma, whose life was her work. As a bonus Kelly recorded everything, so I have a video of the first time Gus, at eleven, managed to zip his own jacket. They are both literally shrieking with joy.

Greta came next. Several years earlier she had a thriving career in publishing, but when her husband died, a grief set in that she could not shake. When she was laid off at her job, she knew she needed to get out of the house. Which is how a sophisticated woman with postgraduate degrees ended up trainspotting daily with my son.

Greta was Swiss, which meant that she was the only person in Gus's life, other than John, who knew the correct order of the stuffed animals on his bed and kept a wary eye in case one was out of place. She called Gus Bert, and he called her Ernie, and I think she genuinely adored him. She loved John, too; he reminded her of her late (British) husband. John had spent many years singing professionally in Germany, so they would speak German together and reminisce, since John's motto is "Everything was better in Germany." Her husband had died of the same heart condition John had, which I wished she would have mentioned a little less often. But she was wonderful. Even if she disappeared every now and then, and could be a tad spacey at other times, I was so grateful for her time with us.

One day, she told me she was taking a road trip with her nephew, and they would spread her husband's ashes in one of his favorite spots in California. She would be back in three weeks, she said. Three weeks turned into a month, and a month into two months. I told Gus she would be back, even as I hired another wonderful woman, Michelle, to accompany him on his adventures. Greta then wrote to tell me that she had a terrible kidney infection, and was staying out in

Los Angeles for treatment. I have a sneaky little email program on my computer that can tell me where an email originates from. Greta was back in New York City. That's when I knew we would never see her again.

A few days later I threw out my back. I was avoiding going to the doctor, as I always do, yet was still desperate for pain relief. I had kept my parents' painkillers, OxyContin and Percocet, from the final months of their lives, just for emergencies like this.

The bottles were all there in my cabinet, lined up neatly. The orderliness of the bottles should have tipped me off. I opened them one by one. There had been scores of pills in there, and now there were none. Unless Henry had decided to become his middle school's drug dealer, I had a good idea where they went. I thought back on Greta's "spaciness" and her disappearances, too, and realized that her time with us lasted as long as there were pills in these bottles. I hope so much there is less pain in her life now.

Every time a caretaker left, I thought Gus would be upset. He was not. At first I wondered if he was cold or uncaring, or just didn't really notice the particulars of who was accompanying him, as long as there was someone. But I don't think that was it. It was just that he thought about friends differently. It was fine if they lived in his computer or in his mind. It was fine if they disappeared only to show up years later. There was no recrimination, only delight at seeing them again.

I want him to understand what real friends are. It's the guy you go to the movies (or bus stop) with, the person you talk to about your annoying parents, the one who shares his beliefs that lightning is scary but sunsets are magnificent. I also want him to know that there is push and pull, that a friend is someone you can have a fight with who still comes back after you finish shouting at each other. A friend is not just lines of text. It's not just the *declaration* that someone is a friend that makes it so.

But let's be fair here. In the age of social media, the idea of friendship is changing for all of us. I have 1806 "friends" on Facebook; tomorrow it will probably be 1809. This is a modest number by the standards of most of my friends. Or are they "friends" with air quotes? I don't know anymore. Now when I go to a party, I often find strangers saying to me, "Oh, I think we're friends on Facebook"—which actually *is* a connection of some sort. It is an immediate point of reference, and if all else fails, we will realize we were both privy to some FB discussion, and X is so smart, and Y is an idiot, and then we're off and running. If that cyber connection means something to so many of us, who am I to define for Gus what friendship is and is not anymore? But if I can't say exactly what friendship is, I have to at least try to give him an idea of what it isn't.

* * *

One afternoon I get a phone call from Gus's school with the words no parent wants to hear: "We are concerned." Mr. T., the school counselor, found that someone had been texting

Gus "inappropriately"—another word that strikes fear into a mother's heart—and that apparently Gus was planning a rendezvous with that person. This is an abbreviated version of the text conversation:

: OK . . . I am on El Roblar by the Farmer & the Cook. This is Samantha btw.
K . . . on my way.
GUS: Oh that's good
SAMANTHA: OK Chi Baby has landed. Should I drop my shit off at the top or grab it later?
GUS: No
SAMANTHA: No . . . just park and come in?
GUS: Yes
SAMANTHA: K. They are checking me in.
Welcome to the Hotel California . . .
GUS: Oh I see
That's cool
SAMANTHA: Is it OK for me to take my stuff to my "dwelling"? They are asking if u are here. . . .
GUS: I am not here. Sure it is ok.
[Me, reading this: *WTF?*]
SAMANTHA: Staff wants to know your ETA . . .
GUS: I don't know what it is
SAMANTHA: I am just the messenger.
GUS: I am in New York City
SAMANTHA: ha, Great . . . then am I the teacher then this weekend?

GUS: No

SAMANTHA: Enjoying this. . . .

GUS: Yes I am

SAMANTHA: Mind fuck a doodle do

GUS: I hate you

SAMANTHA: Fuck yeah!

GUS: You are a villain.

SAMANTHA: Takes one to know one . . .

GUS: I know you're a villain.

SAMANTHA: Are u sure?

GUS: I am sure you are a villain.

SAMANTHA: Wow I'm surprised u took me on. . . . U seem like a very confident man though. . . . I think.

GUS: I am a confident man

SAMANTHA: Yep . . . and intuitive cuz u are totally pushing my buttons . . . well done.

GUS: Thanks.

SAMANTHA: Applause applause

GUS: Yay

SAMANTHA: Mind fuck. . . . cuz u are reminding me of someone. Well done.

GUS: I am not.

SAMANTHA: U are!

GUS: Stop texting me. You are a villain and saying the f word

SAMANTHA: The word is fuck.

My bad!

GUS: I am sorry that I said a bad word
SAMANTHA: No you're not.
Besides fuck is a great word.
I did say a bad word
So what?
GUS: I just don't like it when you say bad words

Apparently Gus is planning a rendezvous in California with a woman named Samantha because he is a playa. Only none of this is true. I dial Samantha's number, and say, essentially, *What the hell.* Samantha is at first appalled, and then we are both intrigued. It turns out that Samantha is some sort of spiritual healer, and there is a convention of healers in Ojai, California. The number of the man leading this conference is one number different from Gus's phone number—and thus the mix-up. Samantha was pleased and flattered that this prominent healer seemed to be getting flirty with her, and the flirting became a power play, and suddenly this conference is taking on a whole new meaning.

It never occurred to Gus, obviously, that his first response to her text should have been "Who are you?" Because when your definition of friend is, uh, emerging, why not just answer whoever texts you? I was kind of pleased to see that once she said "fuck," he decided she was a villain; he's never liked cursing. But this could have been anybody. He would tell her where he lived and what he was doing and whether he was alone; if she had asked for his credit card number nicely and he had one, he would have given it to her. If she had said that she loved him, he might have said it, too. Because, until she was a villain, she was a friend.

Will he always be this innocent?

And will he have someone in his life who can open the can?

* * *

When my worry overwhelms me, I think about Barry.

Barry lives in my building. He is a small man with grizzled graying hair and square glasses that are too large for his face. Every morning he heads somewhere, eyes downcast, carrying a briefcase. In the twenty-five years I've lived here, I've never seen him talk to anyone. If you get in the elevator with him, he presses himself against the walls, staying as far away as possible. He lives with his sister, a tiny woman who looks very much like him, also doesn't talk, and has a problem with her balance.

When Gus was six or seven, I noticed something: Barry waved at him. It was the tiniest of waves, more of a twinkling of the tips of his fingers really. I realized this was because everyone was busy respecting Barry's privacy; everyone, that is, except my son. Gus would shout "HI, BARRY" as the little man scurried by. For a couple of years, Barry's eyes would dart in Gus's direction. But then, over time, there was that teeny-tiny wave.

Every year at our apartment building's Christmas party, Barry and his sister would make an appearance. She would steady herself against the brick wall of the lobby. Neither would speak to anyone, but they seemed content amid the hubbub. This year, I gathered my courage and went to speak to them. The truth is, I wasn't sure they *could* speak.

Well, Barry certainly could; he spoke very softly, but perfectly well and intelligently, with a heavy Bronx accent. He worked in some business I didn't understand, but it involved numbers. And the woman I (and everyone else in the building) had thought was his sister was in fact his wife. I found this immensely cheering.

These were the first words Barry spoke to me after twenty-five years: "How's my friend Gus?"

Thirteen

GETTING SOME

Henry walks into my room at two a.m. "I have phimosis," he says.

"No, you don't," I mutter.

"You don't know what phimosis is, do you?" he replies.

"No, but whatever it is, you don't have it. You also don't have necrotizing fasciitis, lymphatic filariasis, or alien hand syndrome."

"Oh my God, what's alien hand syndrome?"

"It doesn't matter. My point is, you don't have it."

Henry acquires diseases late at night when he Googles. Then he needs to discuss his symptoms.

Phimosis, as it turns out, is a tightness of the foreskin that makes it difficult to move it up and down on the head of the penis. I can't possibly imagine how a fourteen-year-old found this out. Anyway, in an adult it can make intercourse painful. It is not diagnosed before the age of fifteen because lots of boys have a tight foreskin, and it loosens with maturity. But tonight Henry is convinced he will never be able to have sex. And by the way, do girls object to foreskins? He needs to know now.

How does a mother delicately tell her son that in the

condition a woman generally sees a man, a foreskin is a nonissue? I didn't. I just wanted to end the conversation. "Henry, if you do have phimosis, you'll get circumcised and that will solve the problem," I say. He stops talking about it and goes away, but I suspect I didn't help him sleep.

With Henry, every day there are new questions about his body—what's OK, what's not OK, and how girls will feel about it. I tell him to stop looking. Specifically, I tell him he has to start getting dressed in the dark. Nothing stops him.

He also doesn't seem to have gotten the memo that there are things you don't discuss with your mother. My friends insist I'm lucky; mention sex, and their own teenage sons plug their ears and la la la themselves out of the room. Somehow I don't feel my good fortune. Far from being Bleecker Street's Sophie Portnoy, checking the underwear of her precious Alex, I want to know nothing. Instead, I hear everything. Many, perhaps most, of Henry's questions are motivated less by curiosity than by his desire to torture me. When he senses my discomfort he goes in for the kill, like a shark smelling blood in the water. "What do I do if my hands sweat?" "How much of my tongue do I put in when I kiss?" "Do girls like a guy who can last a long time? *How* does a guy last a long time?" "What's the average size of a penis? What's small? What's large? *How big is Dad?*" Most sentences begin with "Is it normal . . . ?" As in, "Is it normal to be able to wank twice in a night? How about three times? Asking for a friend." Recently I've thought about getting a pack of Post-its made up that say "IT'S NORMAL" and slapping them over his mouth before he can open it.

Henry is in fact a handsome boy: lean, dirty-blond Dennis the Menace hair, broad shoulders, green eyes; the line I've heard more than once from single girlfriends is "Boy, you're going to have trouble on your hands." But as far as he's concerned, at fourteen he is the dorkiest dork in Dorkville. The cocky little shit of daytime gives way to a midnight self-esteem spiral. "I'm going to be forty years old, living in this room asking my mommy to make me nachos," he'll say. On a much-anticipated school field trip, he was late but wouldn't get out the door; he insisted on packing his new birthday present into his bag. He was about to miss the bus. Finally he jammed it in, then surveyed the bag sadly. "Yeah. You know who *definitely* gets to make out with a girl on the overnight field trip? The guy who brings his own telescope."

If Henry's worries were contained to his own self, that would be one thing. But his free-floating anxiety often comes to rest on his twin brother. "Mom, do you realize you have one fourteen-year-old son who knows *nothing* about sex?" he said one day as Gus stood nearby, playing *Mario Kart* on his DS. "He has a mustache! He has pubes! I barely have those things, yet *he's* the one who knows nothing."

"He's just dark-haired, Henry, so it's more obvious that—"

Henry was getting himself worked up. "Does he know how babies are made?" he shouted. Gus helpfully patted his stomach. "Does he know how the babies get in there? Does he even know what a condom is?"

Given Gus's lack of dexterity, he has about as much chance of putting on a condom as I have of being named principal ballerina for the New York City Ballet. He still can't manage buttons; left on his own, he just pulls the shirt apart like the Incredible Hulk.

But the condom thing was a problem for sometime in the distant future, right? I put these thoughts aside, just like the thoughts of whether or not he should be able to father children. Right now, it would be good if my deeply affectionate son had just the most rudimentary idea about the birds and the bees.

"Mom, you know how Gus is. This isn't going to happen on its own, and Dad won't talk about it. You have to *do* something."

Henry had a point.

* * *

Nobody really thinks she has to teach her children about sex. I mean, not really, not in the way you might have to teach them, say, how to use a credit card (amazing how fast they catch on to *that*). Kids learn the basics of reproduction, what goes where, and then their natural curiosity takes over. They ask a zillion questions, of either you or their idiot friends, and eventually they figure it out. With boys, in particular, the mechanics come first, and the emotional components come later. (Sometimes way later. Like fifty.) But what if one of the hallmarks of a condition is that there is no natural curiosity? Or, rather, that curiosity is limited to a few very limited subjects—train schedules, weather conditions—and reproduction and love are not among them? What then? Do you leave it all (to quote Sky Masterson) to chance and chemistry? Gus didn't even seem all that interested in his own bodily changes. One night, about two years ago, I said to him, "Honey, you can't just stand under the shower. You need soap, and soon you'll need it even more than you do now."

"Why, Mommy?" he asked.

"Because your body is changing," I said.

"Changing into *what*?" he asked, panicked.

"No, it's not changing into anything *else*, it's just growing, and soon you'll get hormones and—"

"What are hormones?"

"They are body chemicals that will make you grow muscles and hair on your body, and, um, other things change, too."

Gus thought for a moment. "So hormones are magic?" he said.

For many years I never gave adolescence and sex a thought, not only because Gus was so babyish, but also because his preoccupations were already so odd that it all seemed kind of amusing. From the time he was a baby, for example, Gus loved feet. I mean really, really loved them. They even had their own gender: women's feet were feeties, and men's were peet-ies. He never did anything overtly sexual, but feet spoke to him—literally. They meowed, or rather he meowed at them. My neighbor and attorney Jen, a Latina glamazon with car-amel skin, perfect pedicures, and size-twelve clodhoppers, would automatically kick off her shoes when she walked into my house, and Gus would start petting her feet. "Do you think we're encouraging him?" I'd ask worriedly, to which Jen would reply, "Who the fuck cares? Look how happy he is."

Eventually Gus learned to confine his admiration to staring at, or merely complimenting, the feet of women he didn't know. But that took a long time. I spent the first ten years of his life dreading sandal season. When he was eight, we were on the subway platform when he knelt in front of a gorgeous Filipino woman sporting three-inch Manolos and flawless peach nails and started mewing. Coolly she looked down and said, "You could at least buy me dinner first."

I hated getting pedicures—in fact, I dislike anyone touching my feet—but Gus was so enthusiastic that I steeled myself and got them. I reasoned that for a child who did not readily pick up new words, the pedicures provided a teachable moment: he might not know the difference between a quarter or a dime, but thanks to his interest in my toes he knew the difference between rose, raspberry, and magenta. I recorded a bedtime conversation we were all having one night, when Henry and Gus were nine.

GUS: Mommy, what are the names of the new babysitters?
ME: Blair and Kelly.
HENRY: I have to go to law school if I want to get a high position in government, right?
ME: Not necessarily, honey, but it may help.
GUS: Are the new babysitters friendly?
HENRY: What kinds of law are there?

GUS: Do the babysitters know how to sing?
HENRY: What kind of law pays the most?
GUS: Do they have nice feeties?

As it turns out, they both had nice feeties, so after several interviews I picked the one who wasn't insane. Though if the nutty, rude one had had a perfect pedicure, there might have been a problem. I figured that loving pretty feet was such a benign fetish that at best, Gus would become an excellent shoe salesman, and at worst he'd have lots of company in the chat rooms.

* * *

But that was then. Now Gus was fourteen. He still likes pretty toes, and nags me if I don't get mine done. But while it wasn't distressing that Gus may have had an unusual predilection, it was very distressing that he seemed to not understand anything about reproduction and sexually transmitted disease, never mind anything about affection and romance. Could I let him be in *high school*—even a high school for other special ed kids—with this degree of ignorance? But I just didn't know how to broach the subject, because when I mentioned it—"Gus, do you know where babies come from?"—he'd say, "They come from mommies," and then continue talking about the weather or sea turtles or whatever happened to be on his mind at that moment.

First, I decided to attend a very well-intentioned lecture about disability and sexuality at Gus's school. There was a great deal of talk about safety—good touch and bad touch, how to say no, and so forth. Implicit in this discussion was the idea that the bigger problem for people on the spectrum is abuse, not simply sex. Also implicit is the idea that being socially awkward is its own form of birth control. The millions of grown men who collect Star Wars action figures may agree, and of course there's an element of truth here. But at a time when there's better sex ed, less shame around sexuality, an Internet that makes connecting and hooking up easy enough for a kid to master, and so very many people on the spectrum, there are also more opportunities. Also, I could talk to Gus about good touch and bad touch till the creepy cows come home and he would still not understand that kissing, holding hands, and touching—let alone the merging of genitalia—have consequences.

I came home from the lecture a bit disheartened, and no more confident about my ability to impart useful knowledge. Some people are natural teachers. I am the opposite of that. On paper I'm OK. But in person, let me discuss a subject with you that you think you're interested in, and by the time I'm finished with you, you will have unlearned everything you know.

Of course, I was concentrating on the sexual aspect of relationships because that seemed far more concrete, and easier, than dealing with my fears for Gus over the emotional aspects. It is far easier to ponder the question "Will my autistic kid have sex with someone?" than "Will my autistic kid have someone to love?"

* * *

At about the time I was mulling over all of this, I saw an extraordinary documentary called *Autism in Love.* The filmmaker, Matt Fuller, followed four adults on the spectrum who were navigating their way through relationships. One was single and yearning for a girlfriend. Another couple were "high-functioning"—employed, independent—yet struggling with what it takes to make a relationship work. And one man was barely verbal, yet a *Jeopardy!* savant; he had been married twenty years to a woman with a slight cognitive impairment but great emotional intelligence. At the time the movie was being filmed, she was dying of ovarian cancer. His wife's death barely registered on his face or in his words. But as the movie unfolded, the toll on the man became incalculable. And yet for all his silent suffering, his resiliency gave me great hope.

"It just seemed to be such a burning question to me," said Fuller when I called him. "If you don't have a fully developed theory of mind, how do you connect romantically? Do you even want to?" In other words, for autistic people who have difficulty conceiving that the person in front of them might have an entirely different set of needs and desires, what can romantic love mean?

I decided to call one of the women in the film, Lindsey Nebeker. You would never have guessed the hyperarticulate Nebeker had been entirely nonverbal before she was five years old. Nor would you guess this sensual, bohemian woman had all sorts of sensory issues that made it perhaps a little miraculous that she and her husband—who turned his childhood obsession with weather into a job as a meteorologist—were able to touch, never mind make love. But they did. And she didn't mind talking about it.

"Each individual has their own pathway of learning relationships and sexuality. It's kind of tricky," she said. "My father used to say that those of us on the spectrum arrive here without having the antennae that other people have naturally. We notice other people are able to connect their signals and have nonverbal cues, and we have to acquire those tools.

"I really didn't date when I was younger. I had crushes, and friends would tell me if a guy liked me, but I couldn't see it. The signals between me and other people never matched up. I never knew if someone liked me, and I guess I didn't let people know if I liked them." Despite her beauty—which in high school is usually all that's necessary—encounters were few and far between. At the same time, the one person she trusted to talk to about her disability in high school, one of her teachers, ended up sexually abusing her. "I didn't really get it at the time that that's what it was. But I knew it felt wrong."

What was innate for most people required book-learning for Lindsey. She studied Dale Carnegie's *How to Win Friends and Influence People.* The very idea that you had to make other people feel you were interested in them in order to connect to them was completely novel to this young woman. But that realization changed her life.

Still, connecting with people, as much as she yearned for

it, was supremely difficult. As she grew older she was more and more able to appear "regular" to the outside world. But that didn't mean she could have a relationship with a "regular" guy. "Eye contact has always been a problem for me, and it still is. Sometimes, if I'm just in a conversation with someone, it's easier to look at another focal point, in order to concentrate. A face can be very distracting." Another reason had to do with emotional intimacy. "I find that if I have emotions for someone, it makes me feel like I'm made of glass. Like someone can see right through me, can see my exact emotions. That sense of being out of control can make me break down." Which of course is true for all of us—figuratively. Imagine if you *literally* thought someone could see every thought and feeling you had.

With her now husband, it wasn't love at first sight. They met at an autism conference. "I knew when I met him there was something different about him. I thought perhaps an interesting friendship would evolve out of it. I couldn't really label it. At that point in my life, I had vowed to not be in a relationship again." Interestingly, Lindsey was more open to sex than she was to a committed relationship.

Lindsey explains that over the years, sex can be easier than touch. "Our sensory wiring can be advantageous in the bedroom," Lindsey says, laughing. "But at the same time, if there's any kind of resentment or fight between us, touch can be very irritating to me. Even just a touch on the shoulder. Communicating all this is . . . Well, it's still a challenge." Like it is for all of us? "Yes! But maybe more for us, because it's all conscious. We may want to say what we're interested in, but find it very hard to verbalize."

I have no idea whether Gus would find it hard or easy to say what he wants. But did Lindsey have any advice for me about teaching someone who never asks questions about sex?

There is a long silence as she thinks.

"There are many things that took me much longer to figure out than the average person. He may not need to hear from you quite yet. Or he may be aware of many more things than you think."

Lindsey, the woman who was considered profoundly autistic until she was five, reminds me of a notion that's popular in autistic circles. Autism is characterized by developmental delays, but "delay" does not mean "never." It means delay.

* * *

I decided to give this topic a rest for a couple of months. Henry, however, did not.

"I want to show you something, Mom. Gus doesn't even watch porn. It's not normal."

I remembered opening my phone when Henry was seven years old to find that the opening page was something called JuggWorld. Henry, who has always been a terrible speller, nevertheless knew how to spell "boobs" and type it into Google. When I asked him why he wasn't playing *Club Penguin* like he told me, he looked at me solemnly and said, "I'm very interested in the human body."

"First of all," I say, "I don't want to know what you watch. But I need to tell you—"

"Yes, I know, real women sag. I know, I know."

"Second, Gus may not be on the same trajectory as you are, but that doesn't mean . . ."

As I begin my lecture, Henry goes to Gus's computer history, and I see a tiny flicker of alarm in Gus's eyes. This is because, I reason, I am always telling him to watch things that are more age-appropriate than what he actually watches.

"Look!" Henry continues. "Wiggles, *Sesame Street*, *Teletubbies*, Boomerang, and—whoa."

Gus slams down the computer. We will draw the curtain over what we found, but suffice it to say that it put to rest my question about whether he is gay. Also, it appears that someday he may move to Japan. As startling as that is, I immediately don my Autism Mother Goggles, which allow me to see many things that might be a little upsetting in a neurotypical child as progress. *Hey, maybe he won't be watching* Barney *when he's forty. This is OK!*

Then another thing happened.

* * *

However much I love my son, I could not imagine that any girl, anywhere, would find him interesting at this point. Especially a girl like Parker. Parker was slim and leggy, with cascades of wavy brunette hair and eyes the color of blueberries. She was exceedingly pretty, a year older and a head taller than Gus, and given to wearing Star Wars paraphernalia. She wasn't autistic, but had some sort of unspecified learning issues that partially manifested themselves as a need to chat constantly. This was A-OK with Gus, who, while fully verbal, is not exactly a scintillating conversationalist. So a person who filled in the chat gaps, all the time, even chat that involved a fifteen-minute discourse on the importance of protein . . . Perfecto!

They met at school, and then the planning for their "hangout" began. I realize now that the planning was entirely Parker's doing, but Gus would report back to me what she had decided. "Parker is coming over Saturday." "No, she's coming Sunday." "We're staying here." "We're going to a movie." "She's coming next week."

We ended up going to see *The Peanuts Movie* in 3D. I accompanied them; Gus had never gone anywhere by himself. I was happy that he had a new friend. He seemed happy, too. I bought them each a hot dog, chicken nuggets, and the trough of popcorn, but cheaped out after buying one ginormous Diet Coke. I went to sit next to Parker to share the Diet Coke. "Oh, why don't you sit *back there*?" she said, pointing to the row behind. Sheepishly, I went to get my own Diet Coke and settled in behind them. I usually feel queasy at 3D movies, but this one had a beautiful snowfall, and the Red Baron zooming toward me didn't make me want to hurl. I tried to concentrate on the movie, but Gus, occasionally, would look back, putting his hand back to hold mine. Parker would redirect him. "*Eh-eh-EH*," she would say, and dutifully his hands went back to his lap. When we got out of the theater, he automatically reached for my hand, again, and she stopped him, took his hand firmly, and they raced down the street.

After we left the movie, Parker, still hungry, wanted to stop by a diner. Gus almost never eats in a restaurant, but it's not as if he had much choice. Nor could he be his usual picky self about the food. "Gus, eat this lettuce," Parker said. My son, who never touched any vegetable besides avocado (wait, that's a fruit—so I'll stick with "never touched any vegetable"), quickly stuffed it in his mouth. "See? He'll do anything I tell him to," she said, beaming and throwing her arm around his neck. Gus smiled shyly and looked away. Parker readjusted his glasses.

We went back to the house, Parker clasping Gus firmly by the arm. When we walked in, the "hangout" officially began. Armed with a pint of frozen yogurt (a teenager's metabolism is a beautiful thing to see), Parker led Gus to his room and shut the door behind them, pausing only to look back at me with a smile that was—apologetic? A warning? I'm not sure. Then I heard the piano playing, and a great deal of laughter.

"What are you doing here?" said Spencer, my officemate. I had run up three flights of stairs to my office and slammed the door.

"I'm hiding," I said, and explained the situation. "But Gus was just playing piano. He said they were going to play Superhero and Villain. What's the worst that can happen?"

"Besides becoming a grandmother?" Spencer asked helpfully.

I ran downstairs again. This was ridiculous. I shouldn't be the one hiding! Then I remember that Parker had actually dated someone before, a gangly boy who looked like a young Michael Jordan and was a bit of a bowling savant. I knew this because at Gus's last bowling party, while everyone was merrily throwing gutters, this kid got nothing but strikes and spares and didn't seem to think that was unusual. He was fantastic. I'm not sure if he could speak, though. But whatever his issues, he was a fully mature young man.

I just couldn't imagine Parker going from that hunk to a fourteen-year-old with the size and disposition of a boy of nine. She would have certain expectations. Gus was going to be entirely freaked out.

But people find each other. They find their levels.

By the time I got downstairs, Parker and Gus were out of the room. "Is it OK if Parker and I go for a walk, Mommy?" Gus said.

Gus had never been out of the house without another adult accompanying him. Parker's mother had told me her daughter went everywhere by herself, so this was not an issue for her. It would be hard for the average mother to understand the fear of sending a fourteen-year-old out alone. Try this: your kid is fourteen, but then again, when he sees a fire truck whiz by, he becomes three. Would Parker be watching Gus, or was I sending Gus out to an almost certain death?

"Just text me wherever you're going," I said, trying to sound casual.

Gus is, to put it mildly, literal-minded. "I'm in the lobby," he wrote. "Now we're outside the door." "We've walked one block." "We're going to the candy store." (A Parker destination, clearly, since Gus doesn't eat candy.) I held my breath until the two of them settled themselves in the little park directly across from my building. I could see them from my bedroom window. Gus was hopping. "I see you in the window, Mommy!" he texted. "Hi! Hi!"

We waved wildly at each other for a while, and then Parker drew him away and they sat on a patch of grass and talked. I suspect they were talking about Wonder Girl and Ursula, but no matter; they were together.

Henry saw me looking out the window. "Oh my God, Gus is *outside* and we're not there!" he said. Then he noticed that Gus and Parker were holding hands. "Oh, great," he muttered. "My autistic twin has now officially gone further with a girl than I have."

* * *

It is very strange not to know what your autistic child knows, and to not know how responsible you need to be for making sure he knows the basics—including the consequences—about sex.

I am still deeply worried about the idea that he could get someone pregnant and yet could never be a real father. But that is in the future. As I look at my little boy now, I see a person who may never be able to be responsible for another life, but who is nevertheless capable of deep affection, caring, and consideration. Sure, those emotions started with machinery and electronics—trains, buses, iPods, computers—and particularly with Siri, a loving friend who would never hurt him. But he may be ready for humans sooner than I think. Even if the social norms of the rest of the world don't always apply.

Gus has only seen Parker outside of school one other time, but they are together in class and at lunch every day. I had to convince him to go to the school dance; once there, Parker grabbed him, threw her arms around him, and got him on the dance floor. At my suggestion he gave her a huge Hershey's Kiss on Valentine's Day—I told him this would make her happy—and he was very pleased with himself for pleasing *her*.

Until a few days ago he would only say that Parker was his "good friend." But then, late at night, he whispered to me, "I have a crush on Parker."

"That's great, sweetheart. But how do you know?"

"Because she told me," he said.

Fourteen

TOAST

"I think you already know this by heart," I say.

"Just one more time," Henry pleads.

I don't know how plans for my death turned into a favorite bedtime story, but fine. Whatever it takes to make him go to bed before midnight. "OK," I begin, "so what I want is for you to stuff me—"

"Like in *Psycho*?"

"Well, in *Psycho* the mother was grotesque," I continue. "I never understood why Norman Bates's taxidermy on the birds was so good, and his mother became a shriveled mummy. Anyway, I think nowadays I can be freeze-dried and look great." I have no idea if this is true, but it seems like it should be. "So, first you give any usable organs away. Then, I am freeze-dried so that I look exactly like myself, only better. I'll pick out what I wear beforehand. It all depends how old I am at the time. If it were tomorrow it would be something from the Sundance Catalog, because I can still dream, but that won't work if I'm ninety. Anyway. Then, you prop me up in the corner of your living room where I will be doing something I enjoy. Give me a book or my computer."

". . . or I'll have you squinting into your iPhone where you can play Scrabble for eternity," Henry says.

"That would be fine," I say. "Just make sure I look happy. Like I just came up with 'quetzals' or something."

"I have another plan," Henry says, warming to the subject. "I am going to cremate you. And Dad, too. And then I'm going to put googly eyes on your urn. Because everything is better with googly eyes."

"And what about Dad's urn?"

"Also googly eyes. And matchsticks for eyebrows, turned so they look angry." This is John's expression in real life.

"Well, sure," I say. "But you know, first of all, I prefer being stuffed, and if I'm not stuffed and then you don't give me a big party I'm going to haunt you. Second, Dad wants his ashes spread over the woods in Northumberland where he used to play when he was a child."

"OK, now I'm depressed," he says. "Listen, you're getting the urns with the googly eyes and he's getting the matchsticks. Because I want to be able to tell whoever I'm dating that I want her to meet my parents. Every girl likes that. And then I'll bring her home, and you and Dad will be in the urns . . ."

"I can see why you're doing so well with the girls," I say.

"But it's a joke that never gets old," he says. "To me."

"Right."

"Unlike you and Dad. Who get old. And die. And get googly eyes."

"If I'm not freeze-dried and stuck in a corner, looking really good, you're out of the will," I add.

Gus has wandered into the room and listened to about half this conversation. He has no idea what we're talking about. He just wants to make it all better. He comes over and throws his arms around me.

"Henry, don't worry. Mom and Dad will die, but then they'll be back."

Gus believes this. And not in some sort of metaphysical or spiritual way, either. He just knows we'll be back.

* * *

These last five years have been years of loss. This tends to happen when you have kids at a time in life when others are having grandkids. John's parents of course are long gone. My own parents have died in the past few years, one after the other. They were wonderful people, and I want my children to remember them, and they won't.

Of course, kids are pretty resilient. This was Henry at ten, the night after my mother died:

HENRY: [teary, at bedtime] We're never going to see Grandma again, NEVER. She was awesome, and she was your MOM, and YOU'LL never see her again, and I know she's in heaven and she's with Grandpa and everything, but I'll miss her so much and . . .

ME: What is it, honey? Tell Mom.

HENRY: Um. Do we get her house?

ME: Yes.

HENRY: I can get a trampoline! Yesss!

Most kids are wired to recover. And as the years go by, you find ways to make the people you love live on. You mythologize. My beloved golden retriever, Monty, for example, has a kind of living presence in our home, even though he went before my parents. "How many tennis balls could Monty hold in his mouth, Mom?" Henry will ask, apropos of nothing. "Did he really always greet people at the door carrying your underwear?" And always: "He was really, really dumb, wasn't he, Mom?"

We do what we can to erase or at least reshape the final painful life chapters of the people we love. Very shortly before he died, my father was bouncing in and out of sanity. Sometimes we were having long interesting conversations about the first black president, and sometimes there were raccoons jumping into his bed and stealing his Mallomars. The last time I visited my father we were discussing the news, and then he turned to me and said, "I know what you've been up to. I realize journalists don't make much money, but you don't have to keep selling drugs to support your family." The realization that I had a secret life as a drug dealer made him angry, and he screamed at me to get out of his house. He refused to talk to me anymore, and a few days later he died.

This is not exactly the last conversation one would want to have with one's father, but like most horrible things it eventually struck me as funny, and Henry, too. These days, if I'm complaining about work, Henry will say, "Look on the bright side. You can always go back to being a drug dealer."

My mother was a much sweeter human being than my father and died with all her faculties, which gives us a little less to laugh about: Henry was a year older when she died and could not bring himself to go to her funeral. But now, five years later, he still salutes every time we walk by the rehab place near my home where she spent some of her last months.

And he does get me chortling at old videos. "Look, this is going to be you in a few years," he'll say, showing me a tape of my mother talking so intelligently and thoughtfully about one thing or another, surrounded by utter chaos. She had a bit of a hoarding problem. She could not throw away magazines or newspapers, because she would "get around" to reading them, so she had *New Yorker*s in her bedroom dating back to the early '80s. She had a similar fixation on old batteries. There were buckets of them lying around. "They still have some juice in them," she'd say when I tried to throw them out. "You never know when they'll stop making batteries." Apart from believing old batteries would see us through the Mad Max hellscape she apparently envisioned for the future, my mother was tremendously sunny. For years she refused to do anything about the spiders in her bedroom, claiming that they were good for the environment, and when I looked up one day and screamed, noticing there were hundreds of tiny dots on the ceiling, she beamed. "Babies!" she exclaimed.

The hardest loss, in a way, was not my parents but my aunt Alberta, my mother's sister. There is always one person in the family who is the family rock, the preparer of holidays, the rememberer of every event large and small, and she was it.

When her ovarian cancer had spread everywhere, she went into hospice with a few days to live, and proceeded to be there for about six months. She almost made it to ninety-one. One day toward the end, in the summer of 2015, I crept into her room and watched her. Without opening her eyes, she whispered, "What's new?"

"Oh, not much," I said. "You know, the kids are starting school soon, so I've got a lot to do." Then I said the most preposterous thing I could think of. "Also, Donald Trump is president." At which point her beautiful menthol blue eyes opened very, very wide, and we both laughed and laughed.

Later, when she had fully woken up and we were dancing around the issue of her illness, she said, "I'm sorry, I know it's selfish of me, but I don't want to go yet. There are still so many things I'm interested in."

Now that all the elders are gone, it is Henry I look to when I want to reminisce—mostly because *he* likes to reminisce. But when they were sick and dying, Henry was frightened to come near them. Sometimes, to my everlasting shame, so was I. I have never been a tremendously tactile person. And I could never say "I love you" without a tug of embarrassment. The crepey skin, the whiskers on my mother's chin that she could no longer tend to herself, these filled me with horror. I could pat through the bedclothes. I could not hold hands.

That's why my visiting companion was always Gus.

Gus could do all these things, happily. It never occurred to him to be frightened. If he noticed the smells of rot and ammonia that are so much a part of the last weeks of life, they didn't bother him. It never occurred to him to not reach for a hand or lean in for a squeeze, even if that body in the bed could not squeeze back. Most days, I think about the deficits of not understanding a concept as abstract as death, and of course it *is* a deficit. But I've seen the upside of Gus's cluelessness in every hug, every touch, in the very inability to know that my parents would not be getting out of these beds.

* * *

When Henry was seven or eight, he picked the gift for his father's next birthday: a wheelchair. "Mom, it'll be perfect. We'll wheel him around and his legs won't hurt anymore." This was when John was still getting around perfectly well, accompanying Gus every weekend on his beloved forays to the airport, train stations, Port Authority. At that time Henry just thought medical equipment was cool; he had a hankering for oxygen tanks, and it took a while to convince him we didn't need to have them hanging around the house.

The years have gone by, and John, who avoided doctors his whole life, has been working his way through his ailments. This year: heart valve replacement for aortic stenosis and a deep basal cell carcinoma on his nose requiring a skin graft. He's still holding out on the knee replacements, because even though he's just sailed through major heart surgery, he is convinced that knee surgery will kill him. He is no longer able to take Gus out on jaunts. He hasn't been able to return to England to see his family this year, either. He went to the gym religiously three times a week, and now the visits are less and less frequent. Although he still returns to his apartment every night by subway—because fleeing a menopausal wife and two teenage boys is apparently worth the pain—I'm not sure how much longer he'll be able to do this. There is a slowing. There is a softening.

That core of toughness is still there, though, maddening as it is. That inflexibility, too, even when a little change is entirely in his best interest. His knees are bone-on-bone, and yet he still insisted on taking the subway home when he could barely walk; a cab was out of the question because he might have to talk to the driver. That reminded me . . .

"Hey, did you finally take that test I gave you?" I asked him a few weeks ago.

"I did," John said.

I grabbed the questionnaire before he could change his mind, and as I read it my jaw dropped—because his answers to the questions bore almost no relation to the person I had been married to for twenty-four years. Asked to respond to the statement "It does not upset me if my daily routine is disturbed," John said, "Slightly disagree." This, from a person who needed to head home on the same train, at the same time, every day of his life. I particularly loved his response to "Other people frequently tell me that what I've said is impolite, even though I think it is polite." John had checked the box "Strongly disagree." Earlier that day, he had walked into my office, seen my officemate, Spencer, who had just returned from the barber, and said, "I prefer you with longer hair. It makes you look younger." Several days earlier, when I was venting a little about needing to lose weight, he surveyed me carefully and said, "Your stomach's not too bad. And you're pear-shaped. That's much better for your health than being apple-shaped."

Why, thanks! I spend much of my life silently responding to John's "polite" observations with the thought, *Who asked you?*

Using his dubious answers to score the questionnaire, he was entirely neurotypical. When I filled in the answers to reflect my own observations over twenty-five years, he was up to his bad knees in spectrumness.

* * *

As friends have pointed out, I am perhaps the only idiot who marries someone thirty years older who has less money than she does, but what can I say? I love the guy. Unfortunately, that age gap that seemed simultaneously exciting and comforting when the difference was thirty and sixty seems less so when the difference is now fifties and eighties. Henry engages his father's aging by making constant fun of him to his face. When the subject comes up with me, it's a different story. A few weeks ago, when Henry's beloved Jets continued to do what they do best—lose—I came into his room after the game and saw him sitting in the dark with his head in his hands. "I just want them to be in the playoffs *once* so I can watch that game with Dad before he dies," he said.

And Gus? With him there are no discussions, no questions. Dad couldn't take him out because he is old and has bad knees. Simple. Solipsist that he is, he'd keep asking, though. Or did, until a few months ago. There was a shift. It seemed as if his computer came to his emotional rescue, as it so often does.

When Gus is home, he checks in with me about the weather almost every half hour: "Mommy, today's high is sixty degrees—*ahhhhh!*—with a 20 percent chance of a storm. That means it probably won't happen, right?" We rate the temperature according to how it makes us feel, temperate being "ahhhh," hot being "uggggh," and cold being "*eeeeeeeee*." (Henry likes to do his impression of a TV weatherman for Gus: "This morning will start with a low of *EEEEEEE*, reaching *AHHHHH* by midmorning; tomorrow, a small heat front rolling in from the South will bring *UGGGGHH-HHHH* to New York City and the suburbs . . .")

Gus's go-to site for weather is accuweather.com. But then one day, he began adding info to his weather report.

"Mommy, 'Jogger Attacked, Dragged into Wooded Area in Central Park,'" he said. Then he disappeared.

"What did he just say?" John asked.

He didn't seem to want to discuss the stories, but he wanted me to know. "Mommy, 'New York City Bomb Suspect Pleads Not Guilty to Attempted Murder Charges.' Bye."

I was always curious to see what attracted his attention. At first, the deaths seemed to involve terrible accidents—drowning in floods or getting run over by trains. But eventually they became more personal. Gus may not be ready to talk about mortality with a human like, say, his mother, but maybe his machine gives him what he needs in a way I can't.

"'New York City Man Kept Dead Grandmother in Garbage Bag for Months,'" he said. "Also, today's high will be sixty-eight degrees . . ."

"Wait, Gus. Just wait, don't walk away." Henry and I had been giggling about the Gus Doom Report for days, and I had to make sure I didn't laugh. "That is a horrible story. Why do you think the man did that?"

Gus thought for a moment. "Because he's a villain?"

"Well, there's that, and he might be mentally ill." "Might" seemed a bit of an understatement. "But even if he was a sick guy, he may have loved his grandmother and just couldn't let her go, even after she died. Can you understand someone not wanting to let go of someone they loved after they were gone?"

"Gone?" he said. "You mean when they're dead?"

"Yes, honey, when they're dead."

"People die, and then they shouldn't be with you like *that*," he said. "But . . ." He struggled with the thought. "They come back when you think about them. Then you can keep them."

Yes, you can, darling.

* * *

When I learned I was pregnant, I confess my first thought was not the cheeriest one. It was, *Oh, good! I'll have someone to hold my hand when I die.*

I think Henry will be there, my wonderful impossible boy, making me laugh and think as long as we can talk. But it will be Gus holding my hand.

Fifteen

BYE

Henry and I are watching our new favorite documentary, *Baby Animals In The Wild,* where we turn the sound off and supply the narration. His animals are all Scottish and mine are all elderly Jews, because those are the only accents we can do. "*Laddie, aye, the banks of the River Spey are running this year. I could go for a wee dram with my dinner,*" he says as a mother bear catches salmon for her cubs. "*These bones, they're killing me, they get stuck in my teeth,*" I say. "*And you know how much lox costs these days? In my day, it was a penny for lox and a schmear.*" We can do this for hours, despite no one finding it funny but us.

"Do you think Gus will ever live on his own?" Henry says as an elephant helps her baby get out of a ditch.

I don't immediately notice we've switched gears. "*Oy, dahlink, you're getting a little heavy, maybe it's time to switch to cottage cheese and melon?*" I say to the elephants.

"Mom, seriously, what do you think? You know he wants to live in New York City. How will he afford it?"

Gus walks into the living room. Henry turns to Gus. "Gussie, where are you going to be living when you're twenty?"

"Here."

"How about when you're forty?"

"Um . . . here?"

"Mom!" Henry's getting worked up now. "What if I can't make enough money to keep him here? Are you leaving enough? This apartment is very expensive. I don't care if it's paid for, I saw your maintenance bills."

"Oh, look, baby sloths! *Hanging upside down, this is murder to my paws. The arthritis . . .*"

"Mom, focus."

Money worries are never far from Henry's mind. Perhaps this is natural if you're the kind of person who needed to know how interest rates worked when you were six.

Gus is ignoring Henry, albeit for different reasons. "I am going to live here, and help out Jimmy, Jerry, and Dennis." These are our doormen. "And I will go everywhere by myself someday. Right, Mommy?"

"Honey, we've talked about this."

"I know how to go everywhere by myself," Gus says.

"I know you do. That's not the problem. The problem is that you talk to whoever talks to you first."

"I'm a friendly guy," Gus says adamantly.

"You are friendly to people talking to Jesus through the fillings in their teeth," Henry interjects. "And then you give them your money."

"They need money."

This is a well-worn subject, and because of it I keep promising I'll let Gus walk to school by himself—only six blocks away—and then I break my promise. In fact I picked his new high school, the Cooke Center Academy, partially for its proximity, partially for its excellent job-training program for its graduates, and partially because it is run by the nicest humans on the planet. Gus's teacher is funny and smart and great in the classroom, but before she did this, I have no doubt she was great in her other career, as a plus-size burlesque dancer. I respect people who've had a few different lives. Normally Gus doesn't want to do anything for himself. Walking to school is the one thing he wants to do on his own.

But the bigger question, Henry's question, is mine, every moment of every day. Is my little guy going to make it on his own?

On bad days, I focus on all the things he can't do rather than the things he can. And then I think of all the things he couldn't do five years ago that he can do now. Life becomes a series of "on the one hand/on the other" propositions.

- On the one hand, there are so many little bad habits that still need to be broken. The front of his shirt is not a napkin, for instance. On the other, he can put on clothes the right way around. This is no small feat. It was only when I realized that there was something *consistently* wrong with his spatial perception—it wasn't a matter of chance; he would put his shirt and pants on backward 100 percent of the time—that I came up with the genius solution of saying, "Put them on the wrong way." Now he knows that the way that is wrong to him is right. This is working with hair, too. I tell him to brush his hair back, and he ends up looking like a cockatoo. Then I realized that "brush it forward" means he'll slick it back like James Dean.

I'm not sure we'll ever have success with grown-up shoes, though. One day I was lamenting to Mr. Tabone, the principal of his school, that my son would never learn to tie his own shoes. "You know, Judith, you really have to pick your battles," said Mr. Tabone, pointing to his own feet. Mr. Tabone rocks Velcro, and so will Gus.

Why exactly is it that he can't tie his shoes or use buttons, yet can play the piano with fluidity and grace? There are some things I will never understand. But it may be as simple as this: music matters, and the other things don't.

• On the one hand, there is no sign that he will be able to completely handle his own finances. Maybe I shouldn't be comparing him to his brother, Alex P. Keaton, but the whole notion that Gus will ever be able to pay his bills himself or keep money from the predations of, well, anybody seems absurd. "I will have to manage his money," says Henry firmly. "Which means you'll have to leave me a little more. As a consultant's fee. Right? Mom?"

But on the other hand, Gus no longer thinks that the way you get money is that you go to a bank machine, put in your card, and it comes out. For the longest time, as far as Gus was concerned, money didn't grow on trees, but it did live in ATMs. And why not? Machines have always been very good to him. This was just another instance of their kindness. At least he now knows that you work for money. Everything else about it is a little fuzzy.

• On the one hand, Gus says hello to everyone he meets, asks them where they're going, asks about their kids, whether they want to talk or not. On the other hand, he genuinely wants to connect, even if it's on a superficial level. And sometimes that very superficiality is welcome. I sometimes get emails from neighbors if Gus isn't downstairs waiting to say hello in the evening. As one said to me a few days ago, "There's nothing like being greeted by Gus at the end of a tough day."

• On the one hand, Henry still completely dominates him, and Gus will never cross him or fight back in any way. On the other . . . Well, here, too, technology has come to the rescue. At the beginning of the summer Henry and I were having a major blowout over him going to a week of summer camp. Somewhere around the time he was screaming, "*You hate bugs and no air-conditioning, what makes you think I would like them?,*" I received a text from Gus, who was stationed in the other room, listening to us: "Henry is a fool a idiot he always drives you and Daddy crazy." I was so amused, I was able to ignore Henry's ranting for a while—"*Why don't you just put your money in a pile and burn it?"*—and had time to reflect on what this scene would have been like a couple of years ago. It would go something like this: Henry would start to shout, and then Gus would escalate with his own hysterical screaming for no reason whatsoever since no one was even talking to him. Then he would run into his room and slam the door, and I'd have to follow him to make sure he hadn't passed out—because he would probably also be holding his breath.

As I've realized over the years, autism is a dysfunction of empathy, not a lack of it. In the past Gus would often wildly overreact to any discord going on around him; he was incapable of minding his own business. Again and again this got him into trouble, particularly in school, where he would insert himself into every minor fracas if it involved someone he cared for. "Henry is a fool" may not seem like the most helpful response, but it showed a certain sense of proportion—not to mention a correct reading of the situation. Bonus: he didn't try to make himself faint.

* * *

The government has responded to this increasingly common mental condition by throwing lots of money into research. Every week seems to bring a story about a discovery that may lead to better understanding of the condition. Genes that seem to be implicated, structural differences in the brain.

Different neural connections of autistic people, different microbes in the gut, mitochondrial disease . . . tantalizing clues, but nothing even close to definitive.

Science moves slowly—too slowly for many impatient families living with spectrum kids who only want to know, *What are you going to do to make it better?* For example, in 2015 the National Institutes of Health put $28 million toward the Autism Biomarkers Consortium project, which purported to be able to identify children with autism earlier. That sounds good, except what they were identifying were not quantifiable markers like autoantibodies, immunoglobulin levels, T cell counts, and other biological measures of disease. Instead the NIH decided "face processing," "eye tacking," and social communication are biomarkers to be tested by overnight observation in a hospital with repeated EEGs. (If you think being tethered to electrodes and then ordered to sleep in a strange hospital bed is tough for a neurotypical kid, try doing it with a kid on the spectrum. There aren't enough vanilla Frappuccinos in the world to get us through that.)

Using these biomarkers, the goal for early identification of autism is six months. But approximately 40 percent of children develop autism symptoms after the age of one, and even with the best early intervention, only a very small percentage of kids lose their autism diagnosis. For that reason, research funds spent on early diagnosis have been of questionable value. Far better, I think, to channel resources to prenatal genetic testing and treatments to see if something in the womb environment is causing increased rates of autism. Or spend it on investigating environmental causes. Perhaps most important, in addition to research, spend money on treatment; help autistic people to recognize their potential, whatever it may be. I don't need biomarkers for anxiety, or poor eye contact, or poor spatial perception. You only have to go to Gus's school, sit in a chair, and note that half the kids are staying as far away from you as possible, and the other half are speaking two inches away from your face—and almost none are looking at you. There, NIH, I just saved you $28 million.

* * *

"I remember my dream from last night," John said recently over dinner. "You were hitting Gus over the head with the blunt end of a gun."

"So this may not be the best time to ask you if you think I should let him walk to school himself," I said.

"He cannot," John said.

"He is fourteen."

"He will see a fire truck, wave, and get run over by some other car." This was about the five hundredth time John had said the same thing.

"You know, I don't think he will anymore. I've been following him."

"By the way, did you read that story . . . ?" John began.

"Let me guess," I said. "Leprosy is making a comeback. Or

maybe it's about how I can be cured of some disease if I eat Brazil nuts."

"No," he said, "well, not today. It's about that Simon Baron-Cohen. There was a riot."

It was only a riot in John's fevered imagination. But we'd been talking about the psychologist and autism researcher Baron-Cohen, who was the originator of the ill-fated Autism Spectrum Quotient test I had made John take. The story John had read in some British paper was this: Baron-Cohen was denounced by a group of disabled students for a lecture he'd been giving that suggested that within the next five years there would be a prenatal screening test for autism. The disabled students at Cambridge were outraged. Autism was not something to be cured—or eliminated. Instead, it was something that had to be understood culturally, with a shift toward the idea of neurodiversity. Autism was a natural variation in the human condition, and not one meriting alteration or eradication any more than homosexuality.

Henry walked in at that moment, crunching potato chips so loudly that it took me a while to concentrate on what he was saying rather than on my desire to chuck the chips into the garbage.

"So what do you think?" Henry said. "If you could have had the test and known Gus was going to be autistic, would you have taken it?"

"Yes," I said.

"And if you find out tomorrow there's a cure for his autism, would you get it?"

At this point John looked up from his paper.

"No," I lied.

Actually, I have no idea. Autism is a spectrum, and wherever Gus has landed on that spectrum, he is currently a happy person. I adore him just as he is, and autism is so much a part of his Gus-ness.

But over a lifetime, many people suffer deeply. A 2015 study in the *British Journal of Psychiatry* found that people with so-called high-fuctioning autism are about ten times more likely to commit suicide than those in the general population.

So whether or not I would "cure" Gus if given the chance is impossible to answer. On every autism discussion board and support group in the world, there are fights breaking out between adults with autism who hate the idea of a cure, which insinuates that they are damaged, versus parents who are desperate for that very thing. It is the ASD equivalent of the fight in the deaf community about cochlear implants, or among those with dwarfism about limb lengthening. Why can't there be a community and culture of deafness, or little-people-ness, or spectrumness? Why is "normal" always the goal?

Well, it's not, and it shouldn't be.

But at the same time, it's not for the disabled students at Cambridge to condemn the idea of a cure for autism. If you are on the spectrum and you're at one of the finest universities in the world, you cannot speak for the person alone in a room, forever spinning the bright shiny object.

* * *

Often when I watch Gus now, I hear this refrain from "Puff the Magic Dragon":

A dragon lives forever but not so little boys
Painted wings and giant rings make way for other toys
One gray night it happened, Jackie Paper came no more
And Puff that mighty dragon, he ceased his fearless roar . . .

It's at this point that I get weepy and Gus, having no idea what I'm thinking about, comes over and asks me if I'm OK. Admittedly I think "Puff the Magic Dragon" is the saddest song ever written. But now it confuses me. Am I crying at the idea that Gus is moving on from childhood, or the thought that maybe he won't be able to?

But this much I know: today's grim certitudes give way to tomorrow's cheering possibilities, often at the most unexpected times.

I decide to ask him the question I've been prodding him with for the past year. "I haven't asked you this for a while, but I'm curious . . . do you know what being autistic means?"

Gus lays his head down on his computer and covers his eyes. First, as usual, he tries to change the subject. I press him a little. Finally, without looking at me, he says, "I know I have autism."

"And what does that mean to you?"

"It means there are things that are easier for me than for other people, and things that are harder for me. I know I'm different," he continues, his voice barely above a whisper. "But it's OK."

He raises his head and does what is so difficult for him: he looks me directly in the eyes. It's a beginning.

Then, we're back to our current hot topic: "Can I walk to school?"

* * *

It had rained overnight, and now wet tires crushed the damp and exhausted leaves speckling the street. The air was fresh, and the wind was picking up. "Today is blustery," I said. "Today is a minor key," Gus said. We were both right.

"OK, honey!" I said brightly. "Let's go over what we talked about."

"AccuWeather says the temperature is forty-nine degrees—*eeee*—with a 4 percent chance of thunderstorms tonight . . ."

"Not that," I said. "The other stuff."

"I told you, Mommy, I won't talk to anybody. Unless it's a friend."

Great. That doesn't exactly narrow the field.

"*GUS*. Look. Just for six blocks. No talking to 'friends.' Just go. And text me when you're there?"

"Of course, Mommy."

For the first two months of high school I had walked behind him, which meant he'd turn around to wave at me every thirty seconds. So I ducked behind trees and gates like some sort of very bad cartoon spy. He always turned and waved. One day he stopped and chatted with a large African American stranger. This is my shame: I went bounding after him, and in less than one block I was out of breath, glasses askew, my hair wild, sporting what I like to think of as my gym clothes but which are in fact my pajamas. "Hey, I'm Gus's math teacher," the man said. "Hey! I'm Gus's mother, the middle-aged racist," I didn't say. We shook hands and I explained what we were doing. He laughed, and they continued on their way.

I decided the solution would be to have him walk to school with another kid from his school, taller, older, seemingly self-sufficient. Then one day I saw the boy who would have been Gus's safety net bend down in the middle of the street to tie his shoe—headphones on, seemingly oblivious to oncoming traffic, like one of those sacred cows in India that assumes the right of way.

I gave up on that kid, continued to walk Gus, and we practiced texting when he got there. During this time I hadn't had the heart to "test" him by getting a stranger to approach him. Maybe I should have. Yes, I definitely should have. OK, this was dumb. I can't do it. I decide to call the whole thing off.

"Mommy? I don't want to be late." Gus hates being late. He would have made a fine sidekick to Mussolini.

“Right, right, OK,” I say.

It has to happen sometime, right? A dragon lives forever, but not so little boys.

Gus takes about five steps toward his school, then zooms back. “Mommy, you forgot to me ask me the Question.”

“Oh, wait, I forgot what it is . . . Let me see if I remember . . .” His eyes shine as he waits. I estimate that at this point I have asked the question 5,142 times, once a day for every day of his life. And I always joke that I can't remember what it is. To Gus, this never gets old.

“Wait, I think I remember!” I say. “Are you my sweetheart?”

“Yes! I am your sweetheart.” Then he turns and walks away.

I watch until his little silhouette is lost among the morning commuters. About every tenth step, he hops.

ACKNOWLEDGMENTS

There is only one way to write acknowledgments, and that is while drunk. I love all of you! Well, not you. I'm going to lie down now. Bye.

I'm back. This book really happened because a British stranger saw an article I wrote about my son and emailed me. When I googled, I discovered he had published the letters of Nelson Mandela and also written a book called *Do Ants Have Arseholes?* so I knew he had range. Jon Butler, at Quercus: if it weren't for you it wouldn't have dawned on me to write this book. Thanks, too, to his right hand, Katy Follain, who had to take over the editing of the book because while waiting for me to turn it in, Jon produced an entire human being and took paternity leave, because he lives in a civilized country where you can do that sort of thing.

And then there are the people at HarperCollins. David Hirshey has a history of buying books from me that I don't complete; I am so grateful for his bravery and foolishness on this one. When he left HC, he tossed me with a perfect pitch (that will be my only sports metaphor for 250 pages, I promise) into the extraordinarily caring, smart hands of Gail Winston and her associate editor, Sofia Groopman. There is no good writing without great editing. How did I get so lucky?

Now I understand why Academy Award winners thank their agents. I was blessed to have Suzanne Gluck and Tracy Fisher at William Morris Endeavor. I, like so many others, am terrified of Suzanne, which is why I spent two years ducking her phone calls when she was "checking in" to see how this book was going. Very grateful I can talk to her, for now. I look forward to avoiding her calls on many projects in the future.

Pamela Paul and Jen Szalai at the *New York Times* and Kim Hubbard at *People* gave me a chance to do something I loved—book reviewing—thus giving me a much needed distraction from my own book. Other editors who showed more forbearance than I deserved: Kate Lowenstein, Bob Love, Lea Goldman, Rachel Clark, Olessa Pindak, Danielle Pergament.

Now, Laura Marmor. She is my editor and friend at the *New York Times*. She originally urged me to write the article "To Siri, With Love" based on some dopey three-line status update on Facebook. Laura loves tea towels. I don't know why. She deserves a truckload. So if you like this book, you can do two things. You can buy another copy for a friend, or send a tea towel to Laura Marmor, c/o The New York Times, 620 Eighth Avenue, New York, NY 10018.

I would also like to thank the inventors of online Scrabble and Words with Friends. True, I might have finished this book a year earlier without them, but at least they helped me contain the anxiety that comes with having to think hard about your own children. If I had to choose between a night with Gerard Butler and scoring the word QWERTY, it would be a tough call.

Friends and family (and sometimes the line between them is blurred): Jane Greer, Jen Lupo, Jose Ibietatorremendia, Nigella Lawson, Ann Leary, Julie Klam, Laura Zigman, Annabelle Gurwitch, Sheila Weller, Aimee Lee Ball, Lisa DePaulo, Lewis Friedman, Emlyn Eisenach, Nancy Kalish, Megan Daum, Meg Wolitzer, Ellen Marmur, Steven Weinreb, Cynthia Heller, Elissa Petrini, David Galef, Lindsey Cashman, Michelle Sommerville, Amy Lewis, Laurie Lewis, Nancy Sager. A special shout-out to punstress Michele Farinet, who came up with the book's title.

Andrew Nargowala: For the art, for the reluctant therapy, for the fighting, and for the nights you have made me laugh so loud at my computer screen I've woken up my kids.

But mostly this book is dedicated to educators. Some taught me, but most spent their resources of brain and spirit on my sons. For Gus, Margaret Poggi at LearningSpring and Francis Tabone at Cooke Academy. Michael Goldspiel and David Getz, school principals, you rescued Henry at a critical time in his life with your understanding and humor and amazing sense of this-too-shall-pass perspective on all kids who are behaving like morons. Dimitri Saliani, Frances Schuchman, Keith Torjuson, Marie Southwell, Mary Clancy, Dina Persampire, Clare O'Connell. . . OK, why don't I just trot out the faculty lists at my sons' schools?

Sandra Siegel, Henry and Gus's surrogate grandmother

and a retired teacher herself. Dr. Frank Tedeschi, the deeply insightful psychiatrist and all-around sweetheart every parent wishes she had in her children's lives.

Gus has some special teachers outside the classroom as well. Michelle Acevedo, former Marine sergeant and current dedicated trainspotter; I'd go into battle with her anytime. My cherished friend Peter Bloch, who makes it his mission to share with Gus his love of old trains in old New York. Michael Shaw—iconoclast, leader, sweetest man at Grand Central: he was the first train conductor to let Gus announce his route, despite the fact that he could have gotten into trouble for it; he gave him an actual MTA conductor's cap years ago that is still Gus's favorite possession. Jimmy Boshtraf, Jerry Tarantino, and Dennis Badillo Jr. are three of the doormen in my building. Gus has loved to hang with them since he was little. By putting up with him, they have not only given him a bit of on-the-job training; they have taught lessons in work ethics, courtesy, boundary-setting (I mean, they're doormen), and treating people's foibles with humor that you can't always learn in school.

I know this will be shocking to all of you who are married, but I complain a lot about my husband. I complain here, I complain to friends, and if I had a therapist I'm sure I'd be paying money to complain to her. And yet—OK, nope, there is no "and yet." I will always complain. Boy, does he love his children, though. I can only explain him in terms of dogs. Have you had a dog who is so comically grumpy for no apparent reason that your load in life is lightened just by being around him? That's my husband. J, I love you. Now fix your knees.

Mostly and always, to my parents, Frances and Edmund, and my third parent, Aunt Alberta. I wish every day you could be around to see Gus and Henry now.

RESOURCE LIST

US ORGANIZATIONS

Autism Speaks // autismspeaks.org

Founded in 2005; dedicated to promoting solutions, improving lifestyle for those on the spectrum, fostering understanding and acceptance, and advancing research.

Autism Research Institute // autism.com

Founded in 1967; works to improve the health and well-being of people on the autism spectrum through research and education.

Autism Society // autism-society.org

Founded in 1965; advocates for awareness, research, and appropriate services (schools, facilities) for those with autism and their families.

Autism Research Foundation // theautismresearchfoundation.org

Founded in 1990; supports progressive "brain-based" (neurobiological) research, education, family life, and inclusion programs.

National Autism Association // nationalautismassociation.org

Founded in 2003; focuses on providing support, therapy, and medical services for families in areas of greater need; keeping members of the community up-to-date with the latest information about medical research, education, legislation, therapy trends, and safety; funding research studies; and raising awareness about the autism epidemic.

Doug Flutie Jr. Foundation for Autism // flutiefoundation.org
Founded in 1998 by NFL quarterback Doug Flutie; fund-raises through corporate and individual donations, endorsement promotions; awards grants to nonprofit organizations that provide services, family support, education, advocacy, and recreational opportunities to those with autism.

UK ORGANIZATIONS

National Autistic Society // autism.org.uk
Founded in 1962; UK's largest provider of specialist autism services, information, and support.

Autism Alliance // autism-alliance.org.uk
Supports autistic families and adults in homes; runs specialized schools for children with autism; on the Autism Programme Board and the All Party Parliamentary Group on Autism.

Child Autism UK // childautism.org.uk
Provides services and advice for families of children with autism (short-term skills like toilet training through full-time applied behavior analysis programs) and for teachers of children with autism.

Autism Independent UK // autismuk.com
Increases awareness of autism, works to improve quality of life for those with autism, and provides a safe and happy community where they can live, work, and play.

Treating Autism // treatingautism.org.uk
Organization run by parents of children and adults with autism; works alongside carers, nurses, speech/language therapists, dieticians, teachers, researchers, health professionals, and adults with autism to improve quality of life and learning.

Autism Northern Ireland // autismni.org

Works to provide services for the twenty thousand people affected by autism throughout Northern Ireland.

Scottish Autism // scottishautism.org

Provides support services for children and adults across Scotland with a focus on the improvement of quality of life.

CANADIAN ORGANIZATIONS

Autism Society Canada // autismcanada.org

Founded in 1976; works to advocate for and support Canadians with autism and their families in order for those on the spectrum to have full, happy, and healthy lives.

Autism Speaks Canada // autismspeaks.ca

Founded in 2010; promotes collaboration across communities and medical practices, support of families living with autism; committed to research and services across the country.

Canadian National Autism Foundation // www.cnaf.net

Founded in 2000; promotes education of professionals and the general public, providing information and resources to families of people with autism, funding for Canadian-based research and development, and promotion of national autism awareness.

Autism in Mind (AIM) // autisminmind.org

Founded in 2011; works to bring communities together to accept, understand, and support children with autism in order to provide them with a brighter future.

International Autism Foundation Canada // internationalautism foundation.cfsites.org

Founded in 1994; focuses on reaching out through Canadian

Special Education Missions (CSEM). Focus is on toys, games, and family life/parenting tips for families of children with autism.

Autism Society of British Columbia // autismbc.ca

Works to promote understanding, acceptance, and "full community inclusion" of people with autism within British Columbia.

Autism Society Manitoba // autismmanitoba.com

Dedicated to the promotion of quality of life for people with autism through a network of parents and families and a well-fostered community of professionals.

Autism Society of Newfoundland and Labrador // autism.nf.net

Dedicated to promoting the development of individual, life-long, and community-based supports and services for people with autism, their families, and their caregivers.

Autism Society Ontario // autismontario.com

Leading source of information and referral on autism and one of the largest collective voices representing the autism community, connected through volunteer chapters throughout Ontario.

SITES BY PEOPLE WITH AUTISM

Unstrange Mind by Maxfield Sparrow, http://unstrangemind.com/

The Thinking Person's Guide To Autism, http://www.thinkingautismguide.com/2017/

Autism news and resources: from autistic people, professionals, and parents.

Coping: A Survival Guide for People with Asperger Syndrome // http://www-users.cs.york.ac.uk/~alistair/survival

A transcription of the book by Marc Segar, who died in a traffic accident in 1997.

Wrong Planet // wrongplanet.net
American website with articles and blog posts by and for people with autism, Asperger's, ADHD, etc; topics range from politics to book reviews to polls and discussion pages.

Squag // squag.com
Website with a blog featuring posts written for and by children and young people with autism.

Asperclick // asperclick.com
Forum/blog created by Willow Marsden for people with Asperger's.

James' Diary // http://www.autismeducationtrust.org.uk/the-den/diaries.aspx
Blog run by seventeen-year-old James living in England, diagnosed with Asperger's.

Thinking Person's Guide to Autism // www.thinkingautismguide.com
Articles by people with autism, parents of children with autism, partners of autistic people, and professionals in the autism-care fields.

BOOKS

Look Me in the Eye: My Life with Asperger's, John Elder Robison

NeuroTribes: The Legacy of Autism and the Future of Neurodiversity, Steve Silberman

In a Different Key: The Story of Autism, John Donvan and Caren Zucker

Thinking in Pictures: My Life with Autism, Temple Grandin

Uniquely Human: A Different Way of Seeing Autism, Barry M. Prizant, PhD

Ten Things Every Child with Autism Wishes You Knew, Ellen Notbohm

Autism Spectrum Disorder: The Ultimate Teen Guide, Francis Tabone

The Reason I Jump: The Inner Voice of a Thirteen-Year-Old Boy with Autism, Naoki Higashida

The Real Experts: Readings for Parents of Autistic Children, edited by Michelle Sutton

The Autistic Brain: Thinking Across the Spectrum, Temple Grandin

Beyond the Autistic Plateau: A Parent's Story and Practical Help with Autism, Stephen Pitman

Not Even Wrong: Adventures in Autism, Paul Collins

Love, Tears & Autism, Cecily Paterson

Could It Be That Way: Living with Autism, Michael Braccia

I Know You're in There: Winning Our War Against Autism, Marcia Hinds

Nothing Is Right, Michael Scott Monje Jr.

MY PET ORGANIZATIONS

Cooke Center // cookecenter.org/

Special education school and services in New York City.

LearningSpring // learningspring.org

Elementary through middle school education for ASD kids.

Nordoff-Robbins Center for Music Therapy // http://steinhardt.nyu.edu/music/nordoff

Karmazin Foundation

Works with Autism Speaks; money primarily for medical research and autism treatment and services.